Joyce R. C

Carlos Pointe Beach Club

#523

Living with the West Florida shore

Living with the shore

Series editors
Orrin H. Pilkey, Jr.
William J. Neal

The beaches are moving: the drowning of America's shoreline, *new edition*

Wallace Kaufman and Orrin H. Pilkey, Jr.

Living with the East Florida shore

Orrin H. Pilkey, Jr., Dinesh C. Sharma, Harold R. Wanless, Larry J. Doyle, Orrin H. Pilkey, Sr., William J. Neal, and Barbara L. Gruver

Living with the Alabama-Mississippi shore

Wayne F. Canis, William J. Neal, Orrin H. Pilkey, Sr., and Orrin H. Pilkey, Jr.

Living with the Louisiana shore

Joseph T. Kelley, Alice R. Kelley, Orrin H. Pilkey, Sr., and Albert A. Clark

Living with the Texas shore

Robert A. Morton, Orrin H. Pilkey, Jr., Orrin H. Pilkey, Sr., and William J. Neal

Living with the West Florida shore

Larry J. Doyle
Dinesh C. Sharma
Albert C. Hine
Orrin H. Pilkey, Jr.
William J. Neal
Orrin H. Pilkey, Sr.
David Martin
Daniel F. Belknap

Duke University Press   Durham, North Carolina   1984

The publication of this book was supported by a grant from the Florida Coastal Management Office.

Publication of the various volumes in the Living with the Shore series has been greatly assisted by the following individuals and organizations: the American Conservation Association, an anonymous Texas foundation, the Charleston Natural History Society, the Coastal Zone Management Agency (NOAA), the Geraldine R. Dodge Foundation, the Federal Emergency Management Agency, the George Gund Foundation, the Mobil Oil Corporation, Elizabeth O'Connor, the Sapelo Island Research Foundation, the Sea Grant programs of North Carolina, Florida, Mississippi/Alabama, and New York, The Fund for New Jersey, M. Harvey Weil, and Patrick H. Welder, Jr. The Living with the Shore series is part of the Duke University Program for the Study of Developed Shorelines.

Printed in the United States of America on acid-free paper

Library of Congress Cataloging in Publication Data
Main entry under title:

Living with the West Florida shore.

  Includes bibliographical references and index.
  1. Shore protection—Florida—West Florida.
2. Coastal zone management—Florida—West Florida.
3. Coasts—Florida—West Florida. I. Doyle, Larry J.
TC224.F6L585   1984   333.91′716′09759   84-13611
ISBN 0-8223-0516-X
ISBN 0-8223-0517-8 (pbk.)

Contents

Figures and tables

Tables

Foreword

A rising tide of growth imperils the beaches of Florida. So went the headline of a recent article by *Miami Herald* environmental writer Juanita Greene. A lot of other Florida newspaper headlines over the past 2 decades have said the same thing. Yet nothing seems to halt the relentless push to the sea by Florida builders and Florida dwellers.

Take Panama City Beach, for example. According to Deborah Flack, director of the Florida Division of Beaches and Shores, "You can't talk about destroying Panama City Beach because it already is destroyed." Buildings were jammed up against a beach that was retreating landward, and the inevitable happened. The beach disappeared.

Take Walton County, for example. A couple of years ago developers bulldozed dunes on land they were not even ready to develop, just to prevent application of a new state building control (setback) line.

Take Captiva Island, for example. Acres of mangroves on the lagoon side fell victim to bulldozers operating on a weekend when legal steps to prevent this destruction could not be taken.

Now thousands of buildings ranging from simple beach cottages to 20-story condominiums hug the beautiful West Florida shore. In Pinellas County and to the south, these buildings are often located at elevations of no more than 5 feet, too low for residents to safely ride out a storm. The quality of building is variable. Some structures are built well, but many others will be damaged in the high winds and seas that will accompany the next big storm. Escape from the next storm is not feasible for thousands of inhabitants because most evacuation routes are across drawbridges that cannot be depended upon during power outages and rising seas. A more immediate but nonlife threatening problem is the disappearing beachfront in many communities. Repair of costly seawalls and the pumping up of new beach sand promise to add significantly to community tax bills in coming years.

But the future of living on the west coast of Florida has its bright spots, too. For example, there are some sites along the shoreline with relatively high elevations and rows of protective dunes. Furthermore, wise and informed coastal dwellers can take a number of steps to reduce the hazards they face. Most important, concerned residents can be a major political force in the promotion of a rational coastal zone management policy to help save the beautiful West Florida shore for generations to come.

These are the reasons we wrote this book: to help those who already live on the shore, to aid others who may wish to do so, and to preserve the beauty of the West Florida coastline for our grandchildren.

When we began to think about the person to head up this project, Larry Doyle, professor of marine science at the University of South Florida, was a natural choice. Larry has lived near the shore for a number of years and is widely recognized for his studies of the marine environment, on and offshore, of the Gulf of Mexico. He is the author of many technical papers and several books. Co-author Al Hine, also a USF professor, is a coastal geologist with experience on the coasts of Massachusetts, North Carolina, South Carolina, and Florida. Daniel Belknap, who now teaches geology in Maine, has worked on coastal problems in both Delaware and Florida. Dinesh Sharma is an environmental consultant living in Fort Meyers, Florida. He has long been a highly visible power in the struggle for a sound statewide coastal zone management policy. David Martin is a student in marine science at the University of South Florida. Bill Neal and Orrin Pilkey, Jr. (professors of geology at Grand Valley State Colleges and Duke University, respectively), are editing the state-by-state book series of which this title is one entry. And last but not least, Orrin Pilkey, Sr., is a retired civil engineer, living in Charlottesville, Virginia. The senior Pilkey's interest in shoreline construction problems stems from the time his house was severely damaged when Hurricane Camille struck the Mississippi coast in 1969.

This book is one of some 20 projected volumes in the Living with the Shore series. Eventually there will be a book for each coastal state as well as for Lake Erie and Lake Michigan.

As an umbrella book to the series the Duke Press has reprinted with an updated appendix the classic *The Beaches Are Moving: The Drowning of America's Shoreline* by Wallace Kaufman and Orrin H. Pilkey, Jr. This book covers the basic issues dealt with specifically in the state-by-state books.

The series editors have published with Van Nostrand Reinhold Company a construction guide, *Coastal Design: A Guide for Builders, Planners, and Homeowners* (1983) giving detailed coastal construction principles. The prudent coastal dweller should own both *Coastal Design* and the individual state volume.

A lot of people have helped us produce this book. Larry Doyle was supported by Florida Sea Grant. This is Florida Sea Grant College Program Report #63. The Florida Department of Community Affairs, through a grant to coauthor Dinesh Sharma, provided funds to cover some research and printing costs. We would like to extend special thanks to Dr. Asish Mehta and Dr. T.Y. Chia of the Coastal Engineering and Oceanographical Engineering Department at the University of Florida for providing a great deal of information for the compilation of the hazard profiles of individual counties. Lucille Lehman at the coastal engineering archives secured several hundred reports and documents during our research, and the Jacksonville District Headquarters of the U.S. Army Corps of Engineers was extremely helpful in providing us with hard-to-get reports. County and regional planners provided copies of technical reports and ordinances for our review and research.

The overall coastal book project is an outgrowth of initial support from the National Oceanic and Atmospheric Administration through the Office of Coastal Zone Management. The project was administered through the North Carolina Sea Grant Program. Most recently we have been generously supported by the Federal Emergency Management Agency (FEMA). The FEMA support has enabled us to expand the book into a nationwide series including Lake Erie and Lake Michigan. Without the FEMA support the series would have long since ground to a halt. The technical conclusions presented herein are those of the authors and do not necessarily represent those of the supporting agencies.

We owe a debt of gratitude to many individuals for support, ideas, encouragement, and information. Doris Schroeder has helped us in many ways as Jill-of-all-trades over a span of more than a decade and a dozen books. Doris, along with Ed Harrison, compiled the index for this volume. The original idea for our first coastal book (*How to Live with an Island*, 1972) was that of Pete Chenery, then director of the North Carolina Science and Technology Research Center. Richard Foster of the Federal Coastal Zone Management Agency supported the book project at a critical juncture. Because of his lifelong commitment to land conservation, Richard Pough of the Natural Area Council has been a mainstay in our fund-raising efforts. Myrna Jackson of the Duke Development Office and the President's Associates of Duke University have been most helpful in our search for support.

Mike Robinson, Jane Bullock, and Doug Lash of the Federal Emergency Management Agency have worked hard to help us chart a course through the shifting channels of the federal government. Richard Krimm, Peter Gibson, Dennis Carroll, Jim Collins, Jet Battley, Melita Rodeck, Chris Makris, and many others opened doors, furnished maps and charts, and in many other ways helped us through the Washington maze.

We also received a lot of help from Tallahassee officialdom. We would like to particularly note Jorge Southworth of the Department of Commerce and James Stoutamire of the Department of Environmental Regulation. Along the way we received help and encouragement from many of our fellow geologists. We particularly wish to mention our gratitude to Charles Finkle.

Orrin H. Pilkey, Jr.
William J. Neal
series editors

Living with the West Florida shore

1. A coastal perspective

Florida has more than 8,000 miles of ocean coastline, the longest of any of the lower 48 states. Most of the population is concentrated on or near the coast, and the economy depends largely upon tourism, much of which in turn depends upon the beaches. But Florida's beaches are being loved too much. More and more buildings and seawalls crowd the shore, while the beaches become ever narrower. Some beaches have disappeared altogether, to be replaced by massive seawalls.

Our purpose is to educate those who visit, live on, or own property along the coast about beaches and barrier islands. As in everything else in life, there is a right way and a wrong way to live with our state's beautiful beaches and shoreline. We will emphasize how beaches and barrier islands work as systems, and, in the light of the natural oceanographic processes involved we will examine beach stability, beach erosion, and the effects of beach engineering. We also will discuss island safety with respect to natural processes and the hazards of hurricanes and other storms. Finally, there is a layman's guide to help in homesite evaluation, ways to judge construction of new homes, improvement of older homes and condominiums, and a short introduction to Florida coastal law. If you cannot find out everything you wanted to know about Florida beaches and islands in the pages of this book, perhaps the bibliography in the appendix will lead you in the right direction. This volume addresses the beaches along Florida's west or Gulf coast.

Another volume in this series will treat Florida's Atlantic coast separately.

Barrier islands are long, narrow strips of sand separated from the mainland by lagoons. Along Florida's Gulf Coast, barrier islands are found in two strips: the northern Panhandle coast and the west Gulf Coast (fig. 1.1). Within these strips of barrier island coast are short stretches of mainland beach (the Venice area, for example) where no islands or lagoons exist.

Although most of Florida's shoreline has beaches facing the sea, 2 large sections of coast have natural vegetation to absorb breaking ocean waves. Marsh grass and mangrove grow right to the edge of the sea from the end of St. James Island in Wakulla County to Anclote Key in Pinellas County. This section is called the "Big Bend" area. The Ten Thousand Island coast along the extreme southwest portion of the state in Monroe County has spectacular forests of mangrove trees where the sea meets the land. These 2 regions are entirely different geological and ecological systems and, currently, are subject to fewer development pressures (but growing ones, nevertheless) than are the barrier islands.

Status of the shoreline: why are so many beaches eroding?

Widespread erosion is occurring on many of Florida's beaches, as is true for many other portions of the U.S. coast. In some areas

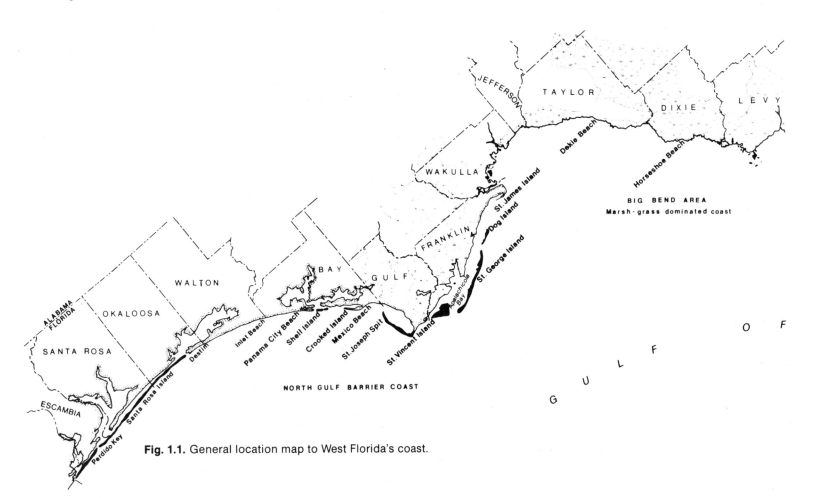

Fig. 1.1. General location map to West Florida's coast.

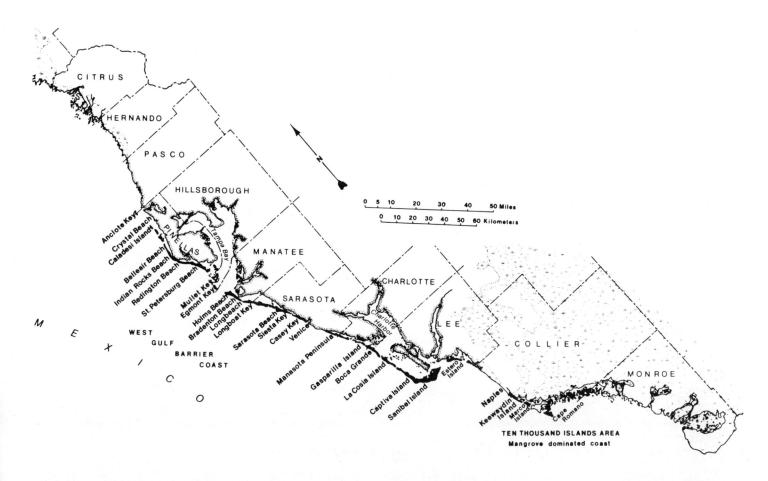

CITRUS

HERNANDO

PASCO

HILLSBOROUGH

N

0 5 10   20   30   40   50 Miles

0 10 20 30 40 50 60 Kilometers

Anclote Keys
Crystal Beach
Caladesi Island

PINELLAS

Tampa Bay

MANATEE

Belleair Beach
Indian Rocks Beach
Redington Beach
St. Petersburg Beach

Mullet Key
Egmont Key
Holms Beach
Bradenton Beach
Longboat Key
Longboat Key
Sarasota Beach
Siesta Key
Casey Key
Venice

SARASOTA

CHARLOTTE

Charlotte
Harbor

LEE

COLLIER

MONROE

M
E
X
I
C
O

WEST
GULF

BARRIER
COAST

Manasota Peninsula
Gasparilla Island
Boca Grande
La Cosia Island
Captiva Island
Sanibel Island

Estero
Island

Naples
Keewaydin
Island
Marco
Island
Cape
Romano

TEN THOUSAND ISLANDS AREA
Mangrove dominated coast

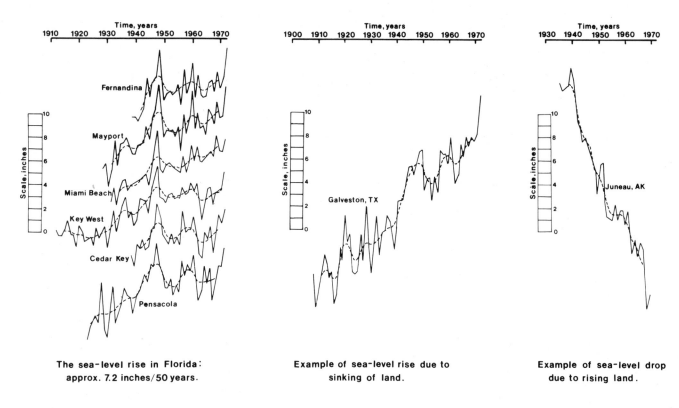

The sea-level rise in Florida: approx. 7.2 inches/50 years.

Example of sea-level rise due to sinking of land.

Example of sea-level drop due to rising land.

Fig. 1.2. Sea-level changes on various shorelines. The sea-level rise in Florida is shown to the left and is typical of most of the American shoreline. The sea level in Galveston, Texas, is rising rapidly because the land is sinking due to groundwater extraction. Off Juneau, Alaska, the sea level seems to be dropping because the land is rising.

the rate of erosion has increased markedly over the last 2 decades.

Examples of shoreline changes in Florida include 1,000 feet of recession of Kice Island since 1885, 350 feet of recession of mid-Manasota Key since 1948, and greater than 250 feet of erosion of St. Vincent Island between 1856 and 1945. Some sections of beaches are actually building out, but these are the exception. Generally, even beaches that are building out, or accreting, do not do so for long. Sometimes the evidence of long-term erosion is indirect but indisputable, such as the loss of the World War II gun emplacements on Egmont Key. Many factors may affect the erosion pattern locally, but the overriding cause is a worldwide rise in sea level. Specific evidence for this rise is provided by tide-gauge records both in the United States and throughout the world. Figure 1.2 shows the sea-level rise over the past 80 years. The upward migration of organisms on a Miami piling is shown in figure 1.3.

Over the past 3 million years or so sea level has fluctuated by as much as 300 feet. The amount of ice in the high latitudes of the earth is the most important single factor controlling worldwide sea level. When the glaciers advance and more water becomes tied up in glaciers, sea level drops. The reverse occurs when the ice melts and retreats. The ice masses in the high latitudes have been slowly shrinking for a long time, and the National Academy of Science recently has warned of the possibility of continued, even accelerated, melting of the ice. The primary cause may be a warming of our atmosphere as a consequence of increasing atmospheric carbon dioxide produced from the burning of fossil fuels—a phenomenon called the greenhouse effect.

Ken Emery at Woods Hole Oceanographic Institute claims that the sea level is currently rising along the American coast at a rate of over 1 foot per century. The Environmental Protection Agency has suggested that a sea-level rise of between 2 and 10 feet by the year 2100 is a real possibility.

The Key West tide gauge shows an increase of only 5 inches over the past 60 years. If the rise continued at that rate, it would amount to 8.3 inches over 100 years. This does not sound like much. In fact, it is often made fun of; but there is another factor involved that takes the fun out. The land lying inland from the coast of Florida, and inland from much of the rest of the east and Gulf coasts, is very low. When sea level goes up, even a small amount, there is a corresponding and much larger horizontal shoreline retreat. Figures 1.4 and 1.5 show how a natural and undeveloped island might respond to a rise in sea level.

Notice how irregular and "jagged" the tide-gauge, sea-level curves are as shown in figure 1.2. There appear to be cycles of about 5 years involving ups and downs ranging from 1 inch to 5 inches. The effect on the beaches of these short-term sea-level fluctuations is not known. However, when we smooth out the jagged peaks on these graphs, we can see that there has been a general rise in sea level since the time the tide gauges were installed. Certainly, the rise over the past 40 to 80 years has been enough to cause a lot of beach retreat, and more should be expected.

Here in Florida we can be thankful that we do not have the rise of 5 feet per century actually measured by tide gauges along portions of the Louisiana coast. The reason that sea-level rise can be different in different coastal areas is that the *land* also may be

Fig. 1.3. Barnacles and oysters have migrated upward on these pilings at Miami, demonstrating a sea-level rise from 1949 to 1981. Photo courtesy of Hal Wanless.

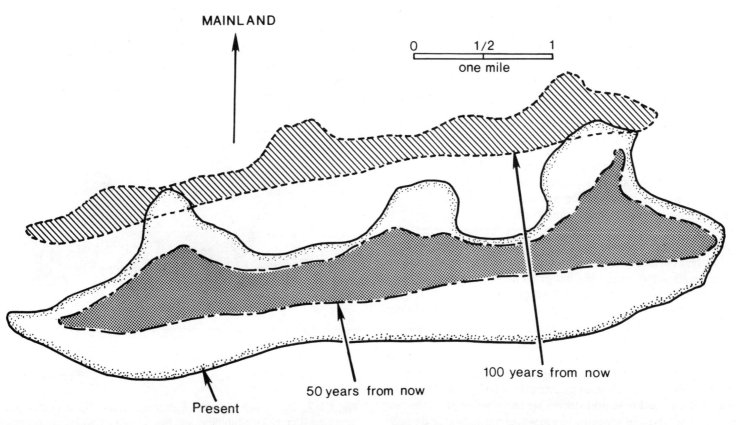

MAINLAND

0   1/2   1
one mile

100 years from now

50 years from now

Present

Fig. 1.4. Theoretically, a very small rise in sea level should produce a very large horizontal shoreline retreat.

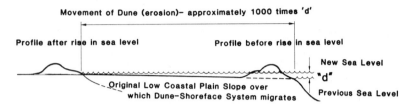

Movement of Dune (erosion)- approximately 1000 times 'd'

Profile after rise in sea level

Profile before rise in sea level

New Sea Level

"d"

Previous Sea Level

Original Low Coastal Plain Slope over which Dune-Shoreface System migrates

Fig. 1.5. A hypothetical example of barrier island migration assuming a 1-foot rise in sea level pushing across the West Florida coastal plain.

sinking or rising. The reason that portions of the Louisiana coast have such a high rate of sea-level rise is that great volumes of Mississippi River muds are compacting and sinking. An opposite and even more spectacular example is the southern Alaskan coast. Here the land is rising much faster than the sea is rising: as a result, the relative change indicates that sea level is falling at a rate of 5.4 feet per century. Florida's coasts are relatively stable compared to these two examples, but given its low elevation any relative rise in sea level is cause for concern.

If the sea level simply rose and no other forces came into play at the shoreline, land areas would gradually flood at a rate that would be totally predictable. As everyone knows, however, the surface of the sea has waves on it, and these waves can become very large during storms. Waves, especially storm waves, can move huge amounts of beach sand in a matter of hours. The higher the sea level, the farther inland storms are likely to reach. In this way the sea-level rise and breaking storm waves combine to cause problems for the Florida shoreline dweller (figs. 1.6, 1.7, 1.8, and 1.9).

Fig. 1.6. A "slab-on-grade" house after Hurricane Eloise (1975) on Panama City Beach. This building was not on pilings and was undermined. Note that the small seawall was ineffective in saving the house.

Fig. 1.7. Beach-front motel after Hurricane Eloise (1975) struck Panama City Beach.

Fig. 1.8. A failed seawall and a destroyed beach on Longboat Key. Photo by Judson Harvey.

Fig. 1.9. Protruding condos on an eroding Pinellas County beach. Photo by Al Hine.

2. Beaches—rivers of sand

Beaches: the dynamic equilibrium

What is a beach?

We all know the answer to this question. A beach is that beautiful ribbon of sand at the seashore that we use for sunbathing and recreation. That is part of the answer, but not all of it. The narrow strip of sand between low and high tides is only a small part of the beach. In reality, the beach is a large and dynamic system most of which is underwater. The beach is shown as a part of figure 2.1. The offshore portion is the zone of active sand movement. Surprisingly, active sand movement may occur at a depth of more than 50 feet and hundreds to thousands of yards seaward of the surf zone. The degree of sand movement is related to the size of the waves striking the beach and the steepness of the underwater section of the beach. What happens on the part of the beach we see depends in large measure upon processes that go on in the offshore portion that we cannot see and that we do not normally think of as part of the beach.

What is the beach made of?

A beach is made of whatever sand and gravel-sized material that happen to be available. For most of Florida the major component of beaches is fine white quartz (silicon dioxide), sand with a mixture of shell and other minerals. In other parts of the world, like the famous Cote d'Azur along the Mediterranean from Cannes to Monte Carlo, the beach may be made of cobbles as big as your fist. Beaches of the Bahamas are composed entirely of shell fragments, pieces of coral, and other calcium carbonate particles. The beaches of Iceland are black volcanic sand. Some beaches of Bermuda are pink because of a kind of organism (*foraminifera*) living in the seaward reefs, and some beaches in Hawaii are green because of a mineral called olivine that has eroded from the lava rock.

Along the west coast of Florida the amount of shell material in beaches is highly variable, but in general the shell content increases from north to south. Some beaches near the mouth of the Apalachicola River have no shells, while some beaches near Naples are sparkling white because of the high shell content. Shells on beaches are the "reason for being" for the tourist industry on many Florida beaches. Most famous of all is Sanibel Island where shell hunting competition is so keen that collectors armed with flashlights scan the beach in the middle of the night.

How do beaches move?

A natural beach is one of the earth's most dynamic environments. But the movement and changes in beaches are not random or accidental. Beaches change according to some very strict natural laws, all of which can be summed up in the dynamic equi-

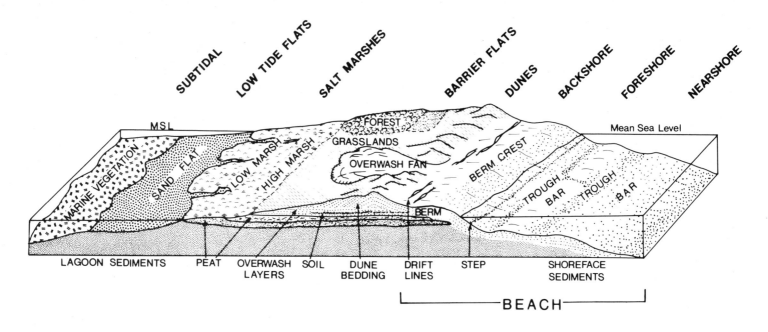

Fig. 2.1. Beach and barrier island natural environment.

librium diagram of figure 2.2. The factors that control beaches are (1) the size of the waves, (2) the rate of sea-level rise, (3) the amount and type of sand, and (4) the shape of the beach. These factors are said to be in a dynamic equilibrium because when one factor changes, the others adjust accordingly to maintain a balance. For example, during a storm when the waves get larger, the beach responds by changing its shape. If fresh sand is pumped onto a beach (beach replenishment), the slope or shape of the beach will vary according to the grain size of the new sand: the finer the sand, the gentler the beach slope.

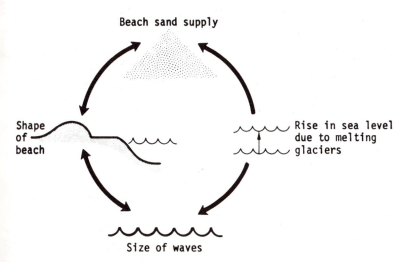

Fig. 2.2. The dynamic equilibrium of a beach.

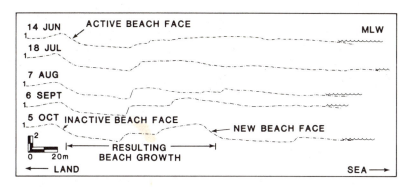

Fig. 2.3. Beach profiles.

Those who spend a lot of time going to one spot on the beach know that it is not the same every time. The offshore bar is in a different place, the beach seems wider or steeper, and the shape itself is often noticeably different. We all have a vague impression that the beach moves and changes shape, but how much? The answer is surprising. Figure 2.3 shows a series of cross sections of a particular beach that grew seaward over 325 feet (99 meters) during one summer, only to be eroded back at least that much during the next winter.

So the beach changes a lot, and huge volumes of sand are in-volved. Geological studies along the Pinellas County coast have estimated that more than 100,000 cubic yards of sand move by any point in either direction up or down the beach in a year's time. This is called the *gross* longshore sand transport. If more sand moves in one direction than the other, the *difference* moving in the preferred direction is called *net* longshore sand transport.

To acquaint you with the amount of sand-moving energy that must be expended on a beach in a year's time, a cubic yard of dry sand weighs 2,150 pounds. Using 100,000 cubic yards as a base, the total amount of sand moved along any point on the Pinellas coast would weigh more than 100,000 tons. The Pinellas coast and all of the sandy shoreline from Anclote Key to Cape Romano are considered to be *low* wave-energy areas, that is, the waves crashing on the beach are relatively small. Coastlines that have much higher waves striking them on a day-to-day basis, such as the east coast of Florida, the North Carolina Outer Banks, or some Pacific Ocean shorelines, have longshore transport rates that are 10 or more times greater.

Sand movement on a beach is very complex. In fact, some scientists and engineers devote much of their professional careers to understanding the physics and mathematics of water motions and sand movement. The beach sand transport system, however, can be simplified into 2 types of directions of sand movement. These are (1) sand moving along the shore, and (2) sand moving offshore and onshore, perpendicular to the shore.

Sometimes when we go swimming at the beach we end up a long way down the beach from where we started. We have been carried there by the longshore current, which obviously is responsible for longshore transport. Such currents result from waves approaching the shore at an angle (fig. 2.4), which causes a portion of the breaking waves' energy to be directed along the beach. The strength of the longshore current rapidly drops off seaward from the breaking waves. Some days when there are at least 3-foot breaking waves, and you can feel the longshore current running along the beach (some people call this undertow), swim beyond the point where the waves are breaking. You will find that you are not carried down the beach nearly so fast as before.

During storms, the longshore currents and turbulence associated with breaking waves can transport large shells, pebbles, bricks, fragments of beach cottages, and even small boulders down the beach. Indeed, during monstrous storms, rocks weighing many

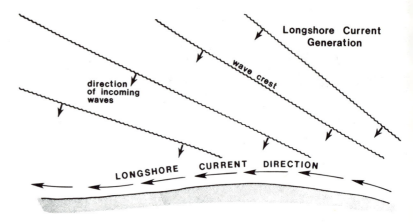

Fig. 2.4. Longshore currents are formed by waves striking the beach at an angle.

tons have been lifted and moved surprising distances. No man-made structure can permanently withstand nature's most powerful storm waves.

Sand transported along the shore may ultimately reach the end of the beach or barrier island. The island may grow laterally like St. Joseph Spit in Wakulla County (fig. 2.5). This is called a recurved spit. More commonly, the sand is deposited into a tidal inlet that separates 2 barrier islands. Good examples of inlets filling in rapidly with sand from longshore transport are the old channels across Captiva Island in Lee County (fig. 2.6) and Stump Pass in Charlotte County (fig. 2.7). Often sand may be transported right

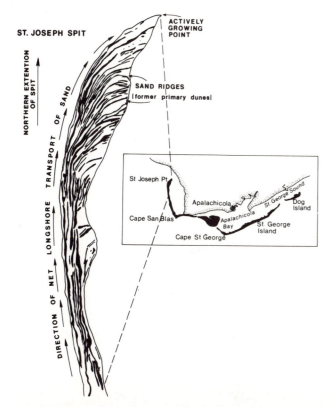

ST. JOSEPH SPIT

ACTIVELY GROWING POINT

NORTHERN EXTENTION OF SPIT

SAND RIDGES (former primary dunes)

LONGSHORE TRANSPORT OF SAND

DIRECTION OF NET

St Joseph Pt

Cape San Blas

Apalachicola

Apalachicola Bay

Cape St George

St George Island

St George Sound

Dog Island

Fig. 2.5. St. Joseph Spit has been formed by buildup of dune ridges paralleling the beach. The sand that forms the ridges is carried to the north by longshore currents in the surf zone.

across an inlet, a transfer from island to island.

Beach sand moves onshore and offshore as well. Gentle waves, more common in the summer, tend to push sand up on the beach. Strong storm waves, more common in the winter or when a hurricane passes nearby, carry sand offshore from the beach. Another way in which sand is removed seaward from the shallow section of the beach is by rip currents (fig. 2.8). These are currents that flow seaward from the surf zone. They can form during both storm and nonstorm conditions and can be hazardous to a swimmer because they are difficult and tiring to swim against.

Where does the beach sand come from?

Ultimately, the quartz sand of beaches is eroded from rocks, such as granite, that compose the continents. Sand grains are carried to the sea by rivers and are mixed with grains of sand formed in the ocean, such as fragments of sea shells. Over a shorter time, sand on a given stretch of beach not only can come from rivers but also may be supplied from the inner continental shelf adjacent to the beach or from nearby eroding cliffs, headlands, and other beaches.

Along Florida's Gulf beaches most of the beach sand has come from the erosion of older geological formations. Some is presently being pushed ashore from offshore. No sand is presently being introduced from rivers. Florida's river mouths have been drowned, or flooded, by the sea-level rise during the last few thousand years. The river's load of sand is deposited far from the coast at the heads of bays and estuaries. Since no sand is being introduced to

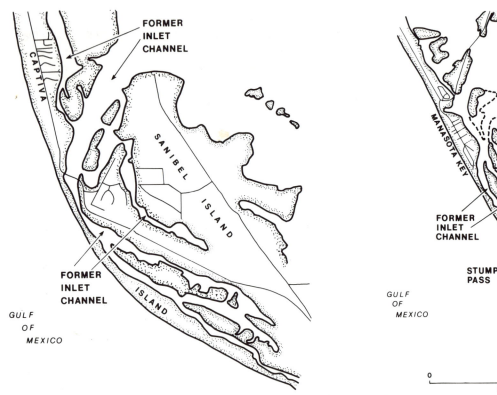

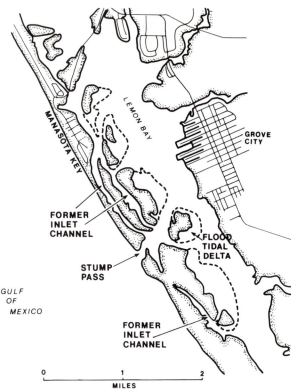

Fig. 2.6. Map showing the position of channels marking former inlets that have since closed up.

Fig. 2.7. Inlet filling at Stump Pass, Charlotte County.

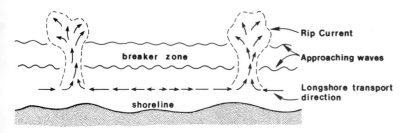

Fig. 2.8. Rip currents that are dangerous to swimmers form when 2 longshore currents meet head-on.

Florida's beaches via rivers, there is a restricted sand supply—only so much to go around. This sand supply problem is a cause of some beach erosion and the reason why some barrier islands are low and narrow. There just is not enough material to make them high and wide. A lack of sand supply is the reason why the barrier island chain ends on the west coast at Cape Romano to the south and Anclote Key to the north. If much more sand had been available when this barrier chain developed, we probably would have more barrier islands to enjoy.

It is important for beach dwellers to know, or at least have some feeling for, the source of sand for their beach. If, for example, there is a lot of longshore sand transport in front of your favorite beach, the beach may well disappear downstream if a wall (groin) is built that blocks this flow. There are literally hundreds of examples of sand-trapping walls on Pinellas County beaches. Actions taken on an adjacent island or inlet will affect your beach, just as your action will affect your coastal neighbors. The severity of these

effects can vary from mild to costly to disastrous (see discussion in chapter 4).

Where do seashells come from?

Most of the seashells found on Gulf Coast beaches are from mollusks that lived on the beach face or just offshore. When these animals die, their shells become sedimentary particles. Many shells wash up on the beach right after storms. Surprisingly enough, however, some of these shells are fossils that are hundreds, thousands, and even millions of years old. In the Venice area the sands are much darker because of black phosphate grains being eroded from a geologic formation (Hawthorn formation) that is millions of years old. Associated with this formation are shells, bones, and sharks' teeth, also millions of years old, that can be found on the beach.

If you use a shell book to identify specimens from a beach, you will find that lagoon shells also can be found on the ocean-side beach. As some islands migrate landward (figs. 1.4 and 1.5), they reexpose the shells that once lived in back-island environments and were buried there. In a few hundred or a thousand years the lagoon shells are thrust up onto the ocean-side beach. But as any beach buff knows, not all beach seashells are fossils by any means. The coquina clam lives in the upper beach and hastens to rebury itself when exposed by sandcastle builders.

How does the beach respond to a storm?

Old-timers and storm survivors from barrier islands have frequently commented on how beautiful, flat, and broad the beach is

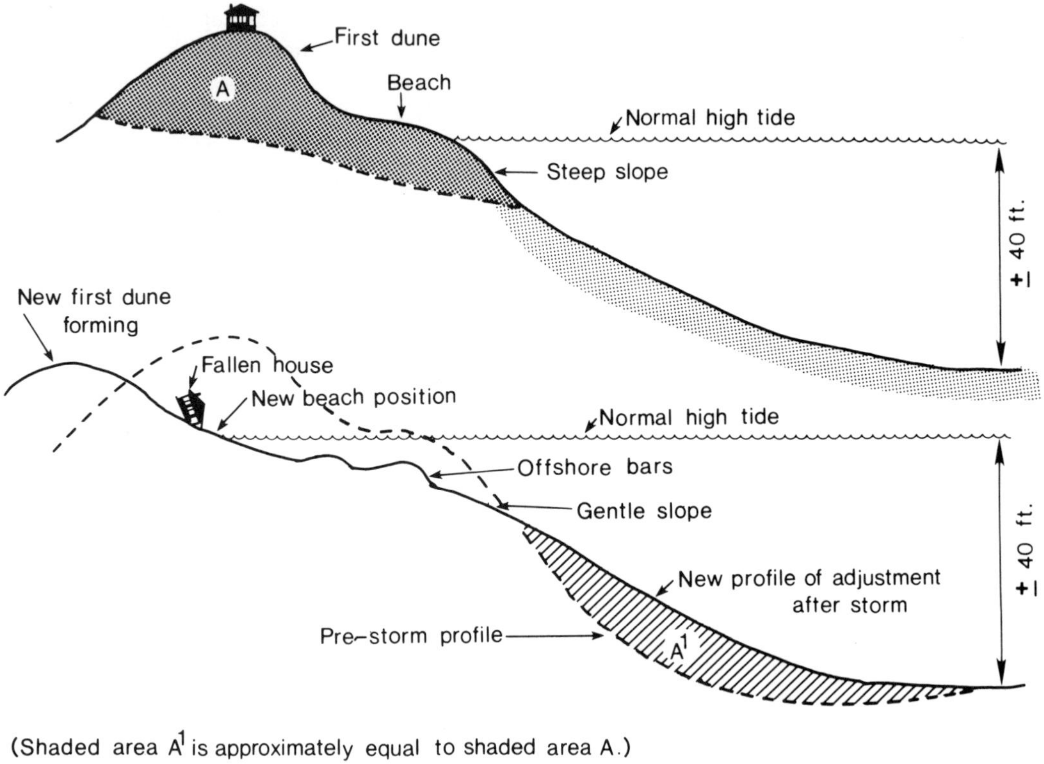

(Shaded area A¹ is approximately equal to shaded area A.)

Fig. 2.9. How beaches respond to a storm. The net result is beach flattening, which allows storm waves to dissipate their immense energy over a broad surface.

after a storm (fig. 2.3). The flat beach can be explained in terms of dynamic equilibrium: as wave energy increases, the storm waves attack the beach and dunes and erode them, moving sand from the upper beach to offshore, thus changing its shape (fig. 2.9). If a hot dog stand or beach cottage or condominium happens to be located on the upper beach or on the first dune, it may disappear along with the sands being transferred to the lower beach. Much of this newly eroded sand is stored in offshore bars. These bars can build up during a storm and can partially protect the beach by causing waves to break on them.

After the storm passes, sand is returned to the beach from the offshore bars by lower energy, "constructional" waves. Along many beaches the sand grains will accumulate on the lower part of the flattened storm profile. The sand builds up and forms a new sand bar, which then begins to migrate shoreward. Eventually, the bar is welded on the exposed portion of the upper beach.

The sand returns to the upper beach, the wind takes over, blowing sand ashore and slowly rebuilding the dunes, storing sand to respond to nature's next storm call. In order for the sand to come back, of course, there cannot be any man-made obstructions such as a seawall between the first dune and the beach. Return of the beach to its prestorm conditions may take weeks, months, or even years.

One final note, if another storm arrives before the beach can fully recover, very severe shoreline retreat is likely to occur. The primary dune line, which may have been only partially attacked during the first storm, may be completely destroyed during this second storm. This one-two punch may deal such a blow to the beach system that it will never fully recover. The sand may be carried far down the beach to be trapped by an inlet; it may be carried far offshore by diffusion or in large rip currents beyond the point where it can easily be returned to the beach, or it may be carried over the beach across the island into the bay or marsh behind. So a period of increased storm frequency can cause permanent beach erosion.

At this point it should be noted that beach *erosion* is a term that only applies to areas where man has built too close to the shore. The natural beach systems frequently retreat after storms or in response to the sea-level rise. However, except for changing its position in space, the beach has the same appearance as before. An eroding beach is not a problem for nature, only for man.

3. Barrier islands

Most coastal plains of the world are fronted by barrier islands. These narrow, flat ribbons of sand are aptly named because they form effective barriers, protecting the mainland shores from the direct onslaught of major storms. Most of West Florida's famous recreational beaches and most of her beach resort communities are on barrier islands. When left alone in their natural state, barriers are dynamic strips of sand capable of withstanding the largest storm nature can throw at them. The islands can even migrate up the coastal plain surface to avoid being drowned by sea-level rise (figs. 1.4 and 1.5).

Each Florida island has a unique shape and demonstrates a unique response to waves, storms, and tides. Each has a unique geologic past. Each barrier or barrier chain must be considered separately and in detail if we are to learn how to live with and on these sandy formations.

To the geologist, Florida's Gulf Coast barriers, particularly the west coast barriers, demonstrate a complexity and variability equal to that of barrier systems anywhere in the world. There is much about barrier islands, their beaches and tidal inlets, that remains to be learned. Surprisingly, the Florida barrier islands are among the least studied in the United States. This is shocking in view of the fact that they have come to be among the most heavily developed in the world. We are developing our coast without nearly enough knowledge and understanding for proper planning, care, and management of this crucial state resource.

To illustrate why and how barrier islands formed, let us begin by dropping sea level by 300 feet or more. This would place Florida's Gulf Coast shoreline between 20 and 100 miles seaward of its present position (fig. 3.1), which is the situation that existed 15,000 or more years ago. Vast glaciers covered the high latitudes of the world, and so much water was tied up in these ice caps that sea level was lowered, exposing most of the present-day continental shelf. When the ice began to melt, sea level began to rise. As the valleys, which were formerly the pathways of the rivers flowing to the sea, were inundated (fig. 3.2), the rising water formed embayments. A glance at a map of today's shoreline shows many such flooded valleys, especially along the Atlantic coast. Tampa Bay and Charlotte Harbor are 2 prominent examples on the Gulf Coast (fig. 1.1).

Longshore currents eroded sand from the stretches of mainland between the bays and deposited it in spits built out into the bays. These spits were cut through during storms, and thus barrier islands were created. Alternatively, the islands may have formed as the rising sea level simply flooded behind and isolated a ridge of sand dunes on a mainland beach. This second method of barrier island genesis (fig. 3.2) is called beach ridge drowning.

Inlet formation by cutting through spits of sand extending into the bay is certainly not an unexpected event. New inlets (often called passes by Floridians) have cut through many of Florida's existing barrier islands in historical storms, and they will cut through islands in the future (figs. 3.3 and 3.4). Between 1883 and 1981 Knight, Don Pedro, and Little Gasparilla islands have had 5 inlets sliced through them. Shell Island has had several inlets in the past, all of which have closed up. Honeymoon and Caladesi islands are separated by the appropriately named Hurricane Pass.

With continued sea-level rise, barrier islands formed one large and continuous island system that migrated landward and up the surface that is now the continental shelf. Needless to say, for the islands to remain as islands, the mainland shoreline had to retreat, too. Viewed in the context of migration all the way across the continental shelf, these narrow strips of sand upon which we build our beach cottages and condominiums were and are indeed dynamic and ephemeral features.

The rate of migration of an island or a shoreline, as common sense would dictate, will be controlled by how fast the sea is rising and by how gentle is the slope of the land. An island migrating across a very gently sloping surface will naturally move faster than an island on a steeper surface. The gradient on some portions of the West Florida continental shelf is extremely flat, and geologists believe that between 15,000 and 5,000 years ago shoreline retreat may have been at times as much as 250 feet per year, or about 5 feet per week. That is rapid beach retreat!

During a time of rapid sea-level rise islands tend to be low, very narrow strips of sand. When the sea-level rise that began 15,000 years ago slowed down 4,000 to 6,000 years ago, some islands actually began to widen. Sanibel Island is an excellent example.

The oldest portions of the barrier islands on the Gulf Coast do not exceed 3,500 years in age, according to Frank Stapor of Exxon Production Research. Many are much younger. Anclote Key, for instance, formed only 1,200 years ago. This might seem like a long time to many people, but from a geologic viewpoint these features are extremely young. The Grand Canyon, for example, is considered a relatively new geologic feature, even though the work of making it has been carried on for *millions* of years.

To complete the story concerning how barrier islands may have formed at the edge of the shelf, we must consider one other mechanism of development. We have seen that barriers can form by 2 methods: (1) cutting off spits that grew into bays and (2) drowning of a sandy ridge on the mainland coast by flooding behind the ridge. The third mechanism is the building up of underwater sand bars. After such bars rise above sea level, vegetation takes hold and a true barrier island is formed. Islands formed this way are generally small and may be easily flooded. Anclote Key in Pinellas County is probably the largest island on the Florida Gulf Coast that may have formed by upbuilding of a submarine bar. This type of barrier is very young. Some may even be forming today. Dog Island built up from a submarine bar and appeared in 1951–52.

It is apparent that most barriers have gone through a complex history of formation and growth. For many islands we will never know their complete history because much of their past has been removed or lies hidden offshore. It is important, however, for us to understand past island history so that we can speculate on how

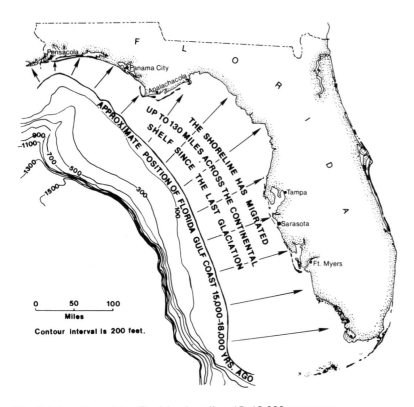

Fig. 3.1. Position of the Florida shoreline 15–18,000 years ago.

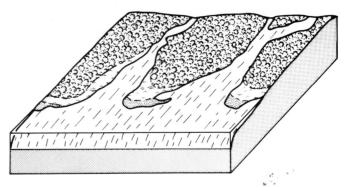

Stage 1: Flooding of river valleys

Fig. 3.2. The origin of barrier islands.

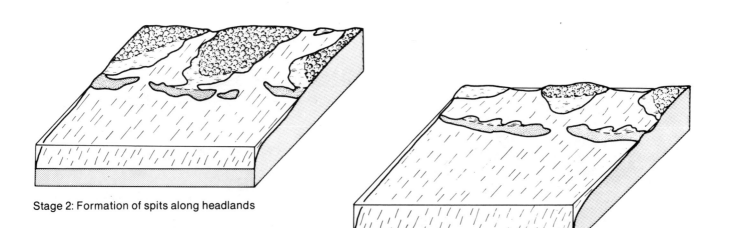

Stage 2: Formation of spits along headlands

Stage 3: Separation of barrier from mainland

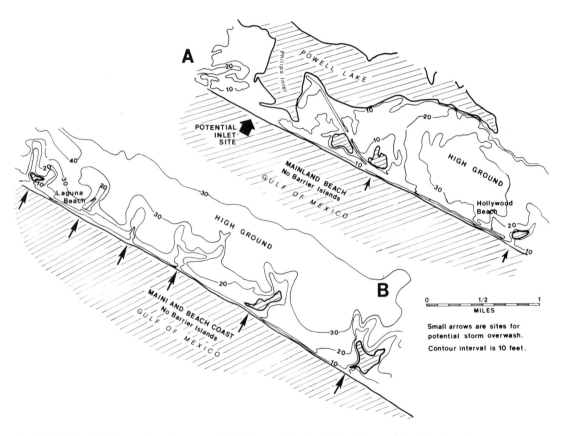

Fig. 3.3. Potential sites of washover and inlet formation for 2 stretches of the Florida shore.

barrier islands may respond to future changes in sea level, wave energy, and sand supply. If we understand the past, we may understand the future.

Barrier islands and inlets

Tidal inlets are highly dynamic and visible factors that can significantly affect and alter the adjacent barrier islands. Very large tidal inlets such as Tampa Bay or Boca Grande Pass (fig. 3.3) do not move great distances in a lateral direction. However, they build large seaward shoals called ebb-tidal deltas. The shoals can extend up to 3 miles offshore and will retreat or bend in response to incoming waves. As a result, the level of wave energy that a nearby beach receives may be focused or dispersed, and the beach will respond accordingly. So, as the offshore shoals change, so do the beaches. In this manner the inlets control the beaches.

Much smaller inlets such as Blind Pass in Pinellas County migrate laterally very rapidly in the direction of net longshore transport. This activity poses obvious hazards (fig. 3.4). If a condominium is constructed on the downdrift side of such an inlet, that structure will soon be threatened by the migrating inlet. Stabilizing efforts are extremely costly and frequently futile. Consequently, construction near all types of inlets should be avoided.

How do barrier islands migrate?

For an island to migrate, the front (ocean) side must move landward by erosion, and the back (lagoon) side must do likewise

Fig. 3.4. Very rapid erosion on a beach adjacent to an inlet. The beach (in Naples, Florida) has retreated past the landward end of the groins. Photo by Al Hine.

by growth. As it moves, the island also must maintain its elevation and bulk. The front side or ocean beach erodes, as we have noted, because storm waves remove sand or because the sand supply to the beach has been stopped. Both of these factors work in concert with continuing sea-level rise (see figs. 1.4, 1.5, and 2.2).

The back side or lagoon shoreline of a barrier island can widen by several mechanisms that include (1) storm washover fan formation and (2) incorporation of flood-tidal deltas from former inlets.

Washover fans are lobes of sand that can extend all the way across a low, narrow island. They form by surges of water flowing across an island during a storm. Where the islands are very narrow and low, they are overwashed frequently, probably at least once a year. During a major hurricane *all* the barriers along Florida's west coast will flood, and all but the widest (the ones that have built seaward during the past several thousand years) will experience overwash in the form of rapidly moving water crossing all or part of the island.

The second way in which sand is added to the back side of barrier islands, thus helping the islands to widen, is through the development of sand shoals called flood-tidal deltas. These sand shoals are formed when sand is carried through an inlet and into a bay by tidal currents; hence, the term tidal deltas. As the inlet migrates away, or if the inlet closes, the old tidal deltas become stabilized by seagrasses, marsh grass, and mangroves and become part of the island. Where there are numerous inlets or where many inlets have formed in the past, this becomes an important process.

In general, where barrier islands are actively migrating and have had a history of numerous inlets opening and closing, flood-tidal deltas form a major portion of the underpinnings of the barrier chain. If it were not for this replenishment (that is, addition of sand to the back side), the islands might have completely eroded away. Figure 3.5 shows the general location of the tidal delta in Cayo Costa Pass. Shoals building up on the tidal delta affect the waves striking Cayo Costa, which in turn affects the shape of the island.

Are the shorelines on the back sides of our islands eroding?

Some back-side shorelines are eroding, while some are not (fig. 3.6). If a healthy salt marsh or mangrove community is growing on the lagoon side, there may be no erosion problem. If a sand bluff or stumps appear on the back side, then beware, erosion is occurring (fig. 3.7). Rates of erosion for most back-island beaches have not been determined.

If most of the ocean-side shorelines of our barrier islands are eroding, what is the long-range future of beach development?

The long-range future of beach development is a function of how individual island communities are able to respond to their eroding beaches. Those communities that choose to protect their ocean-side houses at all costs need only look at portions of the New Jersey and South Florida shores to see the end result. The life span of houses can unquestionably be extended by stabilizing a beach (stopping the erosion; fig. 3.6). The ultimate cost of stopping erosion on a barrier island, however, is loss of the beach. The time required for destruction of the beach is highly variable and depends on the island. Usually a long barrier island seawall will accomplish this destruction in 10 to 30 years, but a single storm can permanently remove a beach in front of a seawall.

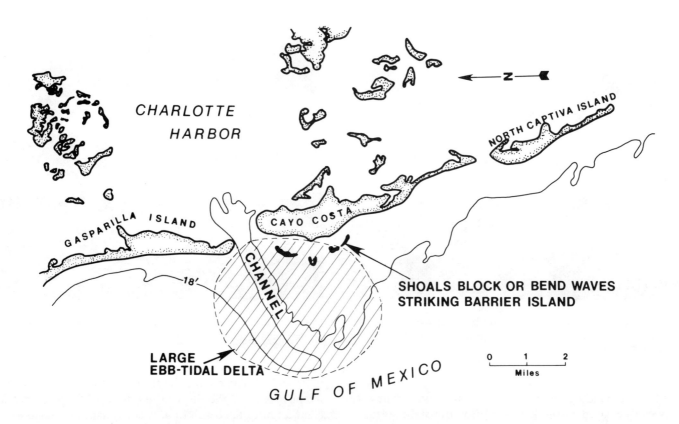

Fig. 3.5. The tidal delta between Gasparilla Island and Cayo Costa.

Fig. 3.6. Revetment to prevent erosion on the bay side of Perdido Key. Very few beaches remain on the bay side of Florida's developed islands. Photo by Dinesh Sharma.

Fig. 3.7. Stumps on an eroding mainland shoreline on Apalachicola Bay. A rising sea level is the likely erosion villain here.

If somehow a community can face facts and buy, move, or let the front row of buildings fall into the sea as their time comes, the beaches can be saved in the long run. Unfortunately, so far in America, the primary factor involved in shoreline decisions (that every beach community must sooner or later make) has been money. Poor communities let the erosion continue. Rich ones attempt to stop it. Needless to say, it is politically impossible to allow a row of tall beach-front condominiums to collapse.

The future of shoreline development in the United States appears to be one of increasing expenditures of money leading to increasing losses of beach. Such a trend in Florida has serious implications both in terms of economics and public safety.

What can I do about my eroding beach?

This is a complex question and is partially answered in chapter 4. If you are talking about an open ocean shoreline, there is nothing you can do unless (1) you are wealthy, (2) the U.S. Army Corps of Engineers steps in, or (3) your community's tax dollars are spent in large amounts.

Your best response, especially from an environmental standpoint, is to move elsewhere. The bottom line in trying to stop open shoreline erosion is that many of the methods employed will ultimately increase the erosion rate. For example, the simple act of hiring a local friendly bulldozer operator to push sand up from the lower beach will steepen the profile and cause the beach to erode more rapidly during the next storm.

Chapter 4 suggests that there are many ways to stop erosion in the short run if lots of money is available, but in the long run erosion cannot be halted except at the cost of losing the beach.

4. Man and the shoreline

Shoreline engineering: stabilizing the unstable

What is shoreline engineering?

Shoreline engineering is a general phrase that refers to any method of changing or altering the natural shoreline system in order to stabilize it. To stabilize means to hold the shoreline in one place. Methods of stabilizing shorelines range from the simple planting of dune grass to the complex emplacement of large seawalls using draglines, cranes, and bulldozers. The benefits of such methods are usually short-lived. Locally, shoreline engineering may actually cause beach retreat, as evidenced by the history of beaches in front of the seawall in Miami Beach (fig. 4.1). Beach erosion influenced by man may be greater and more spectacular than nature's own.

The economic and environmental price of stabilizing the ocean-side beach is stiff indeed. Public awareness of the magnitude of this problem is essential. There are, of course, situations in which stabilization is an economic necessity. Channels leading to our major ports, for example, must be maintained, and sometimes jetties are needed to do so. Without seawalls and jetties, a number of hotels on Florida's west coast would long since have fallen in.

There are 3 major ways by which shorelines are stabilized. These methods are listed below, in decreasing order of environmental safety.

Beach replenishment

If we must "repair" a beach or if we must respond to a retreating shore, beach replenishment is probably the most gentle approach. Replenishment consists of pumping sand onto the beach and building up the former dunes and upper beach. Sufficient money is never available to replenish the entire beach out to a depth of 25 to 50 feet. Thus, only the upper beach is covered with new sand, so a steeper beach is created (fig. 4.2). This new, steepened profile almost always leads to more rapid rates of erosion than those rates that preceded replenishment. Sand is either pumped to the beach from the lagoon, from a pit on the island, from an inlet, or from the shelf. Lagoon sand tends to be too fine and quickly washes off the beach. Furthermore, dredging in the lagoon disturbs the ecosystem, and the hole created affects waves and currents, sometimes harming the back side of the island. Mining the island leaves an unsightly hole of unusable land. As a rule, the best source of sand environmentally—but also the most costly—is the continental shelf. In general, offshore sand is coarser and will stay on the beach longer; at the same time less environmental damage is done by "mining" the shelf. However, Dr. Victor Goldsmith, formerly of the Virginia Institute of Marine Science, warns that when a hole is dug on the shelf for replenishment sand, wave patterns on the adjacent shoreline are likely to be affected. Off the Connecticut

Fig. 4.1. A triple seawall on an eroding shoreline on Longboat Key. The walls were built as the shoreline retreated. Photo by Judson Harvey.

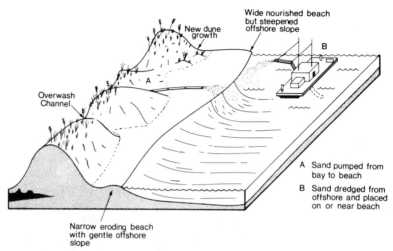

Fig. 4.2. Beach replenishment is the gentlest means of shoreline stabilization, but stabilized beaches tend to erode at a rapid rate.

coast, wave patterns changed by a dredged hole on the shelf quickly caused the replenished beach to disappear.

The west coast of Florida has been the scene of a great number of replenishment projects. Many of the west coast's replenishment projects are smaller than the $68 million, 15-mile face-lifting of Miami Beach and have been carried out by developers, individuals, or communities using locally derived funds. For example, on Marco Island in Florida many small-scale replenishments have been carried out, and on Captiva Island 18 miles of beach was replenished at a cost of $2.8 million, privately raised. Beaches of Pinellas County have been replenished again and again. An excellent approach to replenishment is given by the communities of Estero Island Beach and Keewaydin Island. Channel maintenance of Montanzas Pass and Gordon Pass have furnished sand for the beaches of Estero and Keewaydin, respectively, a decided improvement over the East Florida coast custom of dumping such channel maintenance sand far out at sea.

Beach replenishment, then, upsets the natural system, is costly and temporary, requiring subsequent replenishment projects to remain effective. Historically, if one looks at areas where replenishment has been carried out for a long time (New Jersey, southeast Florida), it is apparent that replenishment is often a way of buying time for increased development. It may not be the intention of the community for such to happen, but often it does. It is important for communities to understand that beach replenishment is strictly a temporary "solution," and a new, broad beach should not be used as justification for a whole new generation of high-rise condos. The bottom line is that replenishment is usually sooner or later replaced by seawalls and groins. The Corps of Engineers refers to beach replenishment as an "ongoing" project, but "eternal" is perhaps a better term.

Nevertheless, beach replenishment is usually less harmful to the total dynamic equilibrium than other methods of responding to erosion. Furthermore, it is a means of stabilization that actually improves the recreational beach rather than causing its ultimate demise, as do the following methods.

Groins and jetties

Groins and jetties are walls built perpendicular to the shoreline. A jetty, often very long (sometimes miles), is intended to keep

Fig. 4.3. Erosion control with groins on Longboat Key. The groins are probably having little effect except to clutter the beach. Photo by Al Hine.

sand from flowing into a channel. Such structures interrupt the flow of sand and may direct sediment offshore, completely out of the beach zone.

Groins, much smaller walls built on straight stretches of beach away from channels and inlets (figs. 4.3 and 4.4), are intended to

Fig. 4.4. Erosion control with groins on Naples Beach. Photo by Dinesh Sharma.

trap sand flowing in the longshore (surf-zone) current. There are groins present today on many Florida beaches. Groins can be made of wood, stone, concrete, steel, or nylon bags filled with sand. Groins also are common in areas where beaches and property are threatened by inlet migration.

Jetties are found at a number of Florida inlets (figs. 4.5 and 4.6).

Fig. 4.5. A small jetty at the end of the beach at St. Petersburg. The jetty has trapped sand and widened the beach. Note the condo adjacent to the inlet, a dangerous location. Photo by Al Hine.

Fig. 4.6. Jetties at Naples, Florida. Sand trapped on the north side causes sand starvation and erosion to the south.

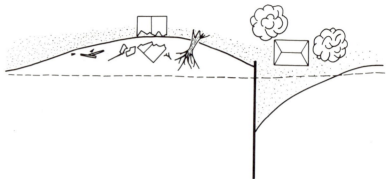

Fig. 4.7. Groins trap sand on one side and cause erosion on the other.

Crooked Island, Shell Island, and Honeymoon Island are northwest Florida examples. The jetties at Doctors Pass and Gordon Pass (near Naples) are more southerly examples. Groins are too numerous to list but are especially thick on Pinellas County beaches as well as on Captiva Island and Manasota Keys. On Sanibel Island only a few groins face the open ocean side of the island, but many more groins (as well as seawalls and revetments) line the back side (lagoon side) of the island.

Both groins and jetties are very successful sand traps. If a groin is working correctly, more sand should be piled up on one side of it than on the other. The problem with the groin is that it traps sand that is flowing to a neighboring beach. Thus, if a groin on a beach is functioning well, it must be causing erosion elsewhere by "starving" another beach (fig. 4.7). The sand trapped by a groin was

Seawalls

Seawalls, built back from and parallel to the shoreline, are designed to receive the full impact of the sea at least once during a tidal cycle. Present in almost every highly developed coastal area, seawalls are fairly common along the West Florida coast.

An even more common type of structure in Florida is the bulkhead, a type of seawall placed farther from the shoreline in front of the first dune—or what was the first dune—and designed to take the impact of storm waves only.

A third member of the seawall family is the revetment (figs. 4.8 and 4.9). These are piles of rock that slope against the first dune (or where the first dune ought to be). They have the advantage of being able to absorb, within rock interstices, some of the waves that crash into them. The water absorbed into the rock cavities reduces the amount of water that rushes back to sea between breaking waves. This also reduces the amount of beach sand lost.

According to the U.S. Army Corps of Engineers, the difference among seawalls, bulkheads, and revetments in terms of damage to the recreational beach is essentially negligible. Building a seawall, bulkhead, or revetment is a very drastic measure on the ocean-side beach, harming the environment in the following ways:

1. They deflect wave energy, ultimately removing the beach and steepening the offshore profile. The time required for this damage to occur is only 1 to 30 years. The steepened offshore profile increases the storm-wave energy striking the shoreline; this in turn worsens erosion and increases the need to build a bigger and better seawall (fig. 4.10).

Fig. 4.8. A revetment near the Bradenton Beach City Hall. At what price have we temporarily halted erosion? Photo by Dinesh Sharma.

going somewhere, and wherever that is will soon be in trouble.

Miami Beach is the nation's most famous example of the long-range results of groin use. After one groin was built (the first after 20 hurricanes), countless others had to be constructed—in self-defense. Prior to the $68 million beach renourishment project of 1977, Miami Beach looked like a military obstacle course; groins obstructed both pedestrian and vehicular traffic. Groins and seawalls had destroyed the beach at Miami Beach.

Fig. 4.9. Revetment on Captiva Island has removed most of the beach. Photo by Dinesh Sharma.

2. They increase the intensity of longshore currents at the base of the wall, hastening removal of the beach (fig. 4.10).
3. They prevent the exchange of sand between dunes and beach. Thus, the beach cannot supply new sand to the dunes on the island, nor can the beach flatten as it tends to do during storms.
4. They concentrate wave and current energy at the ends of the walls, increasing erosion at these points ("the end-around effect").

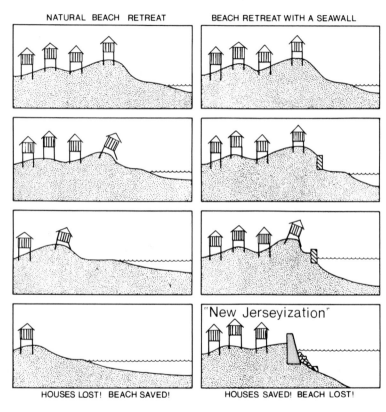

Fig. 4.10. The Seawall Saga.

placed by "bigger and better" and more expensive ones. While a seawall may extend the lives of beach-front structures in normal weather, it cannot protect those on a low-lying barrier island from the havoc wrought by hurricanes, and it cannot prevent overwash or storm-surge flooding. Seawalls may even retard the runoff of flood waters, prolonging the flood.

The long-range effect of seawalls can best be seen in New Jersey, America's oldest developed shoreline. In northern coastal New Jersey, a town building inspector told of the town's seawall history. Pointing to a seawall, he said, "There were once houses and even farms in front of that wall. First we built small seawalls and they were destroyed by the storms that seemed to get bigger and bigger. Now we have come to this huge wall which we hope will hold." The wall he spoke of, adjacent to the highway, was high enough to prevent even a glimpse of the sea beyond. There was no beach in front of it, but remnants of old seawalls, groins, and bulkheads extended for hundreds of yards out to sea.

A philosophy of shoreline conservation

In 1801 Postmaster Ellis Hughes of Cape May, New Jersey, placed the following advertisement in the Philadelphia *Aurora* extolling the beauties of his beach and (with some exaggeration it turns out) Mr. Hughes's facilities.

The subscriber has prepared himself for entertaining company who uses sea bathing and he is accommodated with extensive house room with fish, oysters, crabs, and good

Fig. 4.11. A stretch of shoreline on Casey Key where local shoreline dwellers have been persistent if not successful in combating their erosion problem. In all, 4 different methods have been tried here. Can you spot all 4? Photo by Dinesh Sharma.

Emplacement of the many walls and revetments on Clearwater Beach, Treasure Island, Long Key, and elsewhere must be considered an irreversible act (fig. 4.11). As any casual observer of the Pinellas County beaches knows, the beaches are gone or going in front of the walls, and many of the original walls have been re-

liquors. Care will be taken of gentlemen's horses. Carriages may be driven along the margin of the ocean for miles and the wheels will scarcely make an impression upon the sand. The slope of the shore is so regular that persons may wade a great distance. It is the most delightful spot that citizens can go in the hot season.

This was the first beach advertisement in America and sparked the beginning of the American rush to the shore.

In the next 75 years 6 presidents of the United States vacationed at Cape May. At the time of the Civil War it was certainly the country's most prestigious beach resort. The resort's prestige continued into the twentieth century. In 1908 Henry Ford raced his newest-model cars on Cape May beaches.

Today Cape May is no longer found on anyone's list of great beach resorts. The problem is not that the resort is too old-fashioned, but that no beach remains (fig. 4.12).

The following excerpts are quoted from a 1966 grant application to the federal government from Cape May City. It was written by city officials in an attempt to get funds to build groins to "save the beaches." Though it is possible that its pessimistic tone was exaggerated to enhance the chances of receiving funds, its point is clear:

> Our community is nearly financially insolvent. The economic consequences for beach erosion are depriving all our people of much needed municipal services. . . . The residents of one area of town, Frog Hollow, live in constant fear. The Frog

Fig. 4.12. New Jerseyization!

Hollow area is a 12 block segment of the town which becomes submerged when the tide is merely 1 to 2 feet above normal. The principal reason is that there is no beach fronting on this area . . . . Maps show that blocks have been lost, a boardwalk that has been lost. . . . The stone wall, one mile long, which we erected along the ocean front only five years ago has already begun to crumble from the pounding of the waves since there is little or no beach. . . . We have finally reached a point where we no longer have beaches to erode.

Florida will not have to wait a century and a half for this crisis to reach her shores. Development is here and increasing. Like the original Cape May resort shoreline, stabilization already has affected most of the heavily developed areas on the west coast of Florida. Structures are not placed far enough back from the shore; nor have the builders been so prudent as to place structures behind dunes or on high ground. Consequently, Florida's coastal development is no less vulnerable to the rising sea than was Cape May's, and no shoreline engineering device will prevent its ultimate destruction. The solution lies in recognizing certain "truths" about the shoreline.

Truths of the shoreline

1. There is no erosion problem until a structure is built on a shoreline. Beach erosion is a common, expected event, not a natural disaster. Shoreline erosion in its natural state is not a threat to barrier islands. It is, in fact, an integral part of island evolution (see chapter 3) and the dynamic system of the entire barrier island. When a beach retreats it does not mean that the island is disappearing; the island is migrating. Many undeveloped islands are migrating at surprisingly rapid rates, though only the few investigators who pore over aerial photographs are aware of it. Whether the beach at Fort Pikens National Park on Santa Rosa Island is growing or shrinking does not concern the visiting swimmer, surfer, hiker, or fisherman. It is when man builds a "permanent" structure in this zone of change that a problem develops.

2. Construction by man on the shoreline causes shoreline changes. The sandy beach exists in a delicate balance with sand supply, beach shape, wave energy, and sea-level rise. This is the dynamic equilibrium discussed in chapters 2 and 3. Most construction on or near the shoreline changes this balance and reduces the natural flexibility of the beach (figs. 4.1 through 4.12). The result is change that often threatens man-made structures. Dune removal, which often precedes construction, reduces the sand supply used by the beach to adjust its profile during storms. Beach cottages, even those on stilts, may obstruct the normal sand exchange between the beach and the shelf during storms. Similarly, engineering devices on beaches interrupt or modify the natural cycle.

3. Shoreline engineering protects property on the beach, not the beach itself. Saving buildings is not necessarily a bad thing. But if the shoreline were allowed to migrate naturally over and past the cottages, condominiums, and hot dog stands, the beach would still be there.

4. Shoreline engineering destroys the beach it was intended to save. If this sounds incredible to you, drive to New Jersey or Palm Beach and examine their shore lines (figs. 4.1 through 4.12). Any doubts you may have will disappear when you see the virtual absence of beaches.

5. The cost of saving beach property through shoreline engineering can often be greater than the value of the property to be saved. Price estimates on saving beach property are often unrealistically low in the long run for a variety of reasons. Maintenance, repairs, and replacement costs are typically underestimated because it is erroneously assumed that the big storm, capable of removing an entire beach replenishment project overnight, will somehow by-pass the area. The inevitable hurricane, moreover, is viewed as a catastrophic act of God or a sudden stroke of bad luck for which one cannot plan. The increased potential for damage resulting from shoreline engineering also is ignored in most cost evaluations. In fact, very few shoreline stabilization projects would be funded at all if those controlling the purse strings realized that such "lines of defense" must be perpetual. The fact that shoreline engineering usually is called for on an emergency basis does not help the situation.

6. Once you begin shoreline engineering, you can't stop it! This statement, made by the city manager of a Long Island Sound community, is confirmed by shoreline history throughout the world. Because of the long-range damage caused to the beach it "protects," once engineering starts it must be maintained indefinitely. Failure to allow the sandy shoreline to migrate naturally results in a steepening of the beach profile (down to a depth of 30 or more feet) reduced sand supply, and, therefore, accelerated erosion (see chapters 2 and 3). Thus, once man has installed a shoreline structure to protect the beach, "better"—larger and more expensive—structures must subsequently be installed, only to suffer the same fate as their predecessors (fig. 4.12).

History shows us that there are 2 situations that may terminate shoreline engineering. First, a civilization may fail and no longer build and repair its structures. This was the case with the Romans, who built mighty seawalls. Second, a large storm may destroy a shoreline stabilization system so thoroughly that people decide to throw in the towel. In America, however, such a storm is usually regarded as an engineering challenge and thus results in intensified shoreline stabilization projects.

Questions to ask if shoreline engineering is proposed

When a community is considering some form of shoreline engineering, it is almost invariably done in an atmosphere of crisis. Buildings and commercial interests are threatened, time is short, an expert is brought in, and a solution is proposed. Under such circumstances the right questions are sometimes not asked. The following is a list of questions you might want to bring up if you find yourself a member of such a community.

1. Will the proposed solution to shoreline erosion damage the recreational beach? in 10 years? in 20 years? in 30 years? in 50 years?
2. How much will maintenance of the solution cost in 10 years? in 20 years? in 30 years? in 50 years?
3. If the proposed solution is carried out, what is likely to happen in the next mild hurricane? severe hurricane?
4. What is the erosion rate of the shoreline here during the last 10 years? 20 years?
5. What will the proposed solution do to the beach front along the entire island? Will the solution for one portion of an island create problems for another portion?
6. What will happen if an adjacent inlet migrates or closes up? What will happen if the tidal delta offshore from the adjacent

inlet changes its size and shape or if the channel moves?

7. If the proposed erosion solution is carried out, how will it affect type and density of future beach-front development? Will additional controls on beach-front development be needed at the same time as the solution?

8. What will happen 20 years from now if the inlet nearby is dredged for navigation? if jetties are constructed two inlets away? if seawalls and groins are built on nearby islands?

9. What is the 50- to 100-year environmental and economic prognosis for the proposed solution to erosion if predictions of an accelerating sea-level rise are accurate?

10. If stabilization—for instance a seawall—is permitted here, will this open the door to seawalls elsewhere on the island? (The answer to this one has usually been yes in most other coastal states.)

11. What are the alternatives to the proposed solution to shoreline erosion? Should the threatened buildings be allowed to fall in? Should they be moved? Should tax money be used to move them?

12. What are the long-range environmental and economic costs of the various alternatives from the standpoint of the local property owners? the beach community? the entire island? the citizens of Florida and the rest of the country?

The solutions

1. Design to live with the flexible island environment. Don't fight nature with a "line of defense."

2. Consider all man-made structures near the shoreline temporary.

3. Accept only as a last resort any engineering or stabilization scheme for beach restoration or preservation, and then only for metropolitan areas.

4. Base decisions affecting island development on the welfare of Floridians and the U.S. public as a whole rather than for the minority of shorefront property owners.

5. Let the lighthouse, beach cottage, motel, or hot dog stand fall into the sea when its time comes.

5. Safety on the West Florida coast

Hurricanes

Hurricanes are obviously in a class by themselves as hazards to coastal residents. Danger from these massive storms cannot be overstated. Over the years thousands of people have been killed and billions of dollars lost in property damage. Unfortunately, many more people will die and even larger property losses will result from hurricanes in the future.

Very little can be done to protect property from direct hits by the largest storms, but lives can be saved. The key is to *evacuate early*. Unfortunately, much of the massive development of Florida's west coast has occurred since the last really big hurricane struck. Most residents living on the beaches and barrier islands have never experienced a major hurricane, and many do not take the threat seriously. We recommend that you read the novel *Condominium* by John D. MacDonald, at least the chapters on the hurricane itself, for a vivid description of the destruction and human carnage that could take place in a major storm. MacDonald is especially good at dramatizing all the shortcuts and questionable money-saving practices in building construction and maintenance that are only too plausible. You can believe that such events have happened and will happen again.

To demonstrate that Florida is prone to hurricanes, take a look at figure 5.1 which shows the tracks of some of the hurricanes that have affected Florida's coast. Not only will the beaches be affected in a big storm, but because of the generally low terrain large areas of the mainland will be flooded and should be evacuated as well. For example, planners estimate that in the worst-case scenario (fig. 5.2) of a major hurricane approaching the Tampa Bay region, more than 400,000 people would have to be evacuated—truly a mind-boggling problem to contemplate.

Let's start with basics about hurricanes. A hurricane is a big cyclonic storm that may have a diameter of more than 400 miles and wind speeds of over 74 miles per hour. Big hurricanes pack sustained winds of more than 150 mph with gusts above 200. Typhoons, hurricanes of the Pacific, can have even stronger winds. Hurricanes are often classified in the 5 separate groups shown in table 5.1.

The most dangerous aspect of hurricanes is not wind; it is storm surge. In descending order of danger come waves, flooding due to torrential rains, and finally wind. Storm surge is a mound or ridge of water rising above normal sea level. Only about 10 percent of storm surge is the result of lowered pressure associated with hurricanes. Mostly it is caused by the piling up of water to the right of the wind's direction due to what is known as the Coriolis effect. Storm surge is hardly noticeable in the deep ocean, but when it approaches land, it can come ashore as a devastating wall of water. Bottom topography or shape and the configuration of the coastline also are factors in determining storm-surge heights. Storm surge is in addition to the normal tides. Wind-driven waves

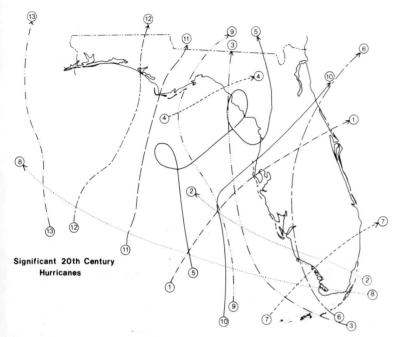

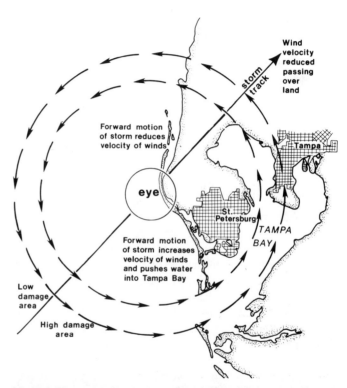

Fig. 5.1. Hurricane tracks off Florida: 1. October 21–31, 1921; 2. September 6–22, 1926; 3. August 31–September 8, 1935; 4. August 31–September 2, 1937; 5. September 1–7, 1950; 6. September 3–13, 1960 (Donna); 7. October 8–17, 1964 (Isabell); 8. August 27–September 13, 1965 (Betsy); 9. June 4–14, 1966 (Alma); 10. October 20–29, 1968 (Gladys); 11. June 1972 (Agnes); 12. September 1975 (Eloise); 13. August–September 1979 (Frederic).

Fig. 5.2. The most dangerous potential hurricane track. If such a storm occurs, a great deal of flooding will result in St. Petersburg and Tampa; hundreds of thousands of people will have to evacuate.

Table 5.1. The Saffir/Simpson Scale with pressure ranges, winds, surge, and damage classifications

Scale number	Central pressures		Winds (mph)	Surge (feet)	Damage
	Millibars	Inches			
1	980	28.94	74–95	4–5	Minimal
2	965–979	28.5–28.91	96–110	6–8	Moderate
3	945–964	27.91–28.47	111–130	9–12	Extensive
4	920–944	27.17–27.88	131–155	13–18	Extreme
5	920	27.17	155+	18+	Catastrophic

ride on top of the storm surge and can actually be magnified by the deepening water. Figure 5.3 shows National Oceanic and Atmospheric Administration (NOAA) predictions for storm surge on the Florida coastline. There are 2 sets of numbers. The smaller numbers give the estimated storm surge for a 100-year storm and the larger figures for a storm like Hurricane Camille (1969), among the worst ever to hit the United States.

Do not be fooled by the use of the term 100-year storm, which means that a storm of this magnitude can be expected to occur at a given point once every 100 years. Several areas in Louisiana and Mississippi experienced 2 100-year floods in 1 month in the spring of 1983. Moreover, if you buy a condominium in a new building with an expected life of 50 years, there is a 50–50 chance the structure in which you live will go through a 100-year storm in its lifetime. Be aware also that emergency planners expect to have to evacuate as many as 250,000 people in the Tampa Bay area for a hurricane with a storm surge of only 4 or 5 feet, and hurricanes of

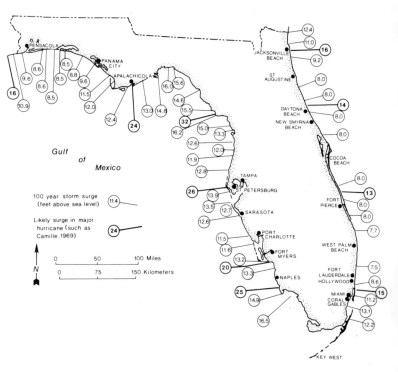

Fig. 5.3. Storm-surge map for Florida. Modeled after "Hurricane Survival," NOAA Coastal Hazards Program.

that size can be expected much more frequently than can the 100-year storms.

The key to survival if a hurricane strikes a barrier island is simply for you not to be there. No building is hurricane-proof. If you live on a barrier island beach or low-lying area, you must be ready to evacuate. Consider it part of life in a shore community, and you may have to do it more than once when the hurricane changes course and misses your area. Failure to evacuate in time can cost you and your family your lives. Do not stay in a high building. Vertical distancing of yourself from the storm is no solution. Few buildings, even well-constructed ones, can withstand the larger storm surges. In addition, some condominiums are insecurely fastened to the pilings on which they rest. Finally and sadly, it is impossible to identify certain buildings in which some shortcuts have been taken that will make them less able to withstand the forces of storm surge.

The high density of population living on the barrier islands and in the low-lying coastal cities makes evacuation a major problem. For instance, a lead time of as much as 18 hours is needed to evacuate the community of St. Petersburg Beach. To evacuate this single small area, 8,000 to 9,000 cars must pass over only 2 low-lying bridges that would be under water 6 to 8 hours before a major hurricane hits. In addition, the access roads leading to the bridges will submerge even earlier. So start early, because stalled cars and submerged roads can block escape. Consider the magnitude of the potential traffic jam if a major evacuation of more than 400,000 people is attempted (fig. 5.3).

A further problem lies in our ability to predict the path of a hurricane. While a storm's general direction is sometimes predictable, specific direction is not, and hurricanes sometimes change direction suddenly. NOAA tracks and predicts the paths of hurricanes. In 1983 they said that they would change the way in which they predict where a hurricane is going. They now issue a warning in probability terms, i.e., there is a 10 percent chance it will hit an area or a 30 percent chance, etc. Local officials are worried that this kind of warning will cause many people to resist evacuation. Don't be one of them. *Get out early*. Consider that to evacuate even a community of modest size like St. Petersburg Beach may take 18 hours, in which time a hurricane moving at 30 miles an hour would travel 540 miles.

Local officials on the West Florida coast are concerned about the potential for disaster. In addition, local TV and radio stations and newspapers each year run public information stories about what to do in the event a hurricane strikes. Pay attention. Call your fire department or local Disaster Preparedness office for more information and plan your evacuation timetable and route. *Be a survivor!*

Smaller storms and normal beach processes

There is no such thing as a perfectly safe place to live on a West Florida barrier island or mainland shoreline. Given the right conditions, hurricanes, floods, wind and wave erosion, and inlet formation can attack any coastal area. Furthermore, human activity such as dune bulldozing or mangrove removal almost always lessens the stability of the natural environment and increases the danger to local dwellers. On the other hand, some areas are much safer than others to live on, and the wise coastal dweller looks for site-safety clues before making a purchase.

The idea is to place your home in the least dynamic zone where it is most likely to survive a storm. For example, a condominium built in the pine forest on the back side of Perdido Island should prove to be a much safer site than the same condominium built on some of the low and narrow stretches of Gasparilla Island. The idea is to identify the rates and intensities of storm activity for a given segment of shoreline and to use this knowledge as a basis for safe selection.

Most individuals who come to dwell near the beaches of West Florida know very little about the environment they have come to enjoy. Often their only advice with regard to site safety comes from realtors and developers who are not known for their objectivity in this regard. Sometimes home or condo purchasers take the attitude that "since thousands of people live on this island, it can't be all that unsafe." To make that assumption when choosing a West Florida homesite is a serious and fundamental error (figs. 5.4 and 5.5).

Clues for the wise and prudent

The following sections might well be entitled "What you've wanted to know about near-the-beach site selection but couldn't figure out who to ask."

Fig. 5.4. Very tall buildings on the very low Marco Island. Photo by Al Hine.

Elevation. The elevation of a site at the coast is its single most important natural attribute; the higher the elevation the better. Common sense tells us that low, flat areas are subject to destructive wave attack, storm overwash, and storm-surge flooding. Remember that storm flooding of low areas often occurs from the lagoon side of an island.

Dunes. The next best thing to being at a high elevation is to

Fig. 5.5. Tall buildings hugging the beach on very low building sites at Naples, Florida. Photo by Al Hine.

Fig. 5.6. This was Panama City Beach after Hurricane Eloise (1975). The dunes were cut back more than 50 feet, but since no seawalls were located here the beach remained broad and healthy. Photo by Bob Morton.

have some rows of dunes between you and the forces of the sea (figs. 5.6 and 5.7). Dunes form by wind blowing sand in shore from the beach. Primary dunes or the first row of dunes are the most important and should be protected at all costs. Examples of a well-developed primary dune system can be seen on St. Joseph Spit in Gulf County. Unfortunately, many if not most primary dunes have disappeared on West Florida's developed islands (fig. 5.8).

Dunes are important because they may block or reduce the impact of storm waves, they increase the elevation of homesites, and in a big storm they furnish a reservoir of sand with which the beach can flatten (see chapter 2). Preserve them if humanly possible.

Vegetation. Type of vegetation is a good indication of stability, age, and elevation of a shoreline area. For example, the pine for-

Fig. 5.7. Artificial dune in front of a condo construction site near Naples, Florida. Artificial dunes do not act like natural dunes. This dune line might better be called a sand dike! Photo by Al Hine.

Fig. 5.8. Dunes were removed from this section of Fort Meyers Beach and replaced by buildings and seawalls. Photo by Dinesh Sharma.

ests along stretches of the Bay County and Walton County shorelines are areas that are inundated relatively rarely. However, the marshy or mangrove shorelines of Dixie and Levy counties in the Big Bend area and Collier and Monroe counties in the Ten Thousand Island area are subject to frequent flooding. Pine forests do not survive frequent saltwater intrusions. Mangroves and salt marsh vegetation, on the other hand, require frequent intrusion.

In general, the higher and thicker the vegetative growth the more stable the site and the safer the area. Since maritime forests (forests adjusted to some salt spray) take many years to develop, the homeowner can be assured that a homesite in such an area has been around a long time and offers a relatively higher degree of safety.

Australian pines flourish on the beaches of West Florida, but

Fig. 5.9. Lovers Key after a June 1982 storm. Australian pines are poor erosion buffers because of their shallow root systems, as can be seen from these stumps. Australian pines also are poor dune builders, and on the shorefront they should be replaced by native vegetation. Photo by Dinesh Sharma.

Fig. 5.10. Mangroves.

unfortunately they are now considered to be unwelcome intruders. They have shallow root systems and are easily blown over or knocked over (fig. 5.9). They do not stabilize dunes or help them to grow, and they furnish large amounts of debris to be cleaned up after beach storms. These pines are best removed and replaced with dune vegetation.

Mangroves used to line the quiet water shorelines of much of West Florida (fig. 5.10). The most famous and extensive mangroves in Florida are those lining the Ten Thousand Islands. Alas, mangroves just like dunes have fallen victim to the bulldozer and

Fig. 5.11. Salt marshes. Photo by Joan Hutton.

are no longer a common sight on many developed Florida islands. Where they remain they should be protected. They are excellent erosion buffers, and in fact their intertwining roots even cause the shoreline to build out.

Salt marshes (fig. 5.11), ecosystems that also flourish in quiet waters, often line the bay sides of islands as well as the mainland margins of bays, especially in the Big Bend area. The true salt marsh is a unique botanical environment because it consists of a single species of plant, *Spartina alterniflora*. These areas are important breeding grounds for many species of marine life and offer considerable protection from wave attack as well. Much of West Florida's salt marsh has been filled in for development (which is now illegal), but new marsh can be planted on bay shorelines, and such replanting should be considered as an erosion buffer alternative to bulkheads and seawalls.

Soil profile. Soils are a good clue to building site stability. White sand overlying yellow sand to a depth of 1 to 3 feet suggests stability because such a profile requires many years to develop. You can spot soil profiles in road cuts, finger canals, or in a self-dug pit. Keep in mind that even formerly stable forested areas can be eroded, so you may find a "stable" soil profile about to fall into the sea. Such is the case along the shoreline from Apalachicola Bay westward where beautifully developed soil profiles are exposed on actively eroding sand bluffs.

Soil that contains shells means the site is subjected to storm overwash. Alternatively, the shells may be there because new sand was pumped in from the bay. Generally, shells that arrived by natural storm processes are the same color as those on the beach, while land artificially pumped up contains a high percentage of black shells. Other things being equal, soil with shells is less desirable than soil without shells.

Finger canals. A very common man-made island alteration in West Florida is the finger canal (fig. 5.12). Finger canal is the term applied to the ditches or channels dug from the lagoon or bay side of an island into the island proper for the purpose of providing

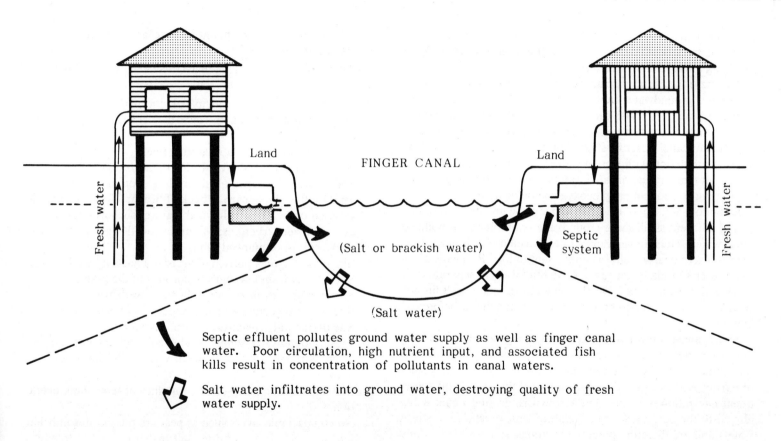

Septic effluent pollutes ground water supply as well as finger canal water. Poor circulation, high nutrient input, and associated fish kills result in concentration of pollutants in canal waters.

Salt water infiltrates into ground water, destroying quality of fresh water supply.

Fig. 5.12. A finger canal system.

everyone with a waterfront lot. Canals can be made by excavation alone, or by a combination of excavation and infill of adjacent low-lying areas (usually marshes).

The major problems associated with finger canals are (1) lowering of the groundwater table, (2) pollution of groundwater by seepage of salt or brackish canal water into the groundwater table, (3) pollution of canal water by septic seepage into the canal, (4) pollution of canal water by stagnation due to lack of tidal flushing or poor circulation with lagoon waters, (5) fish kills generated by the higher temperature of canal waters, and (6) fish kills generated by nutrient overloading and deoxygenation of water.

Bad odors, flotsam of dead fish and algal scum, and contamination of adjacent shellfishing grounds are symptomatic of polluted canal water. Thus, finger canals often become health hazards or simply places near which it is unpleasant to live. Residents along some older Florida finger canals have built walls to separate their cottages from the canal! On the other hand, some well-flushed canals make beautiful homesites. Talk to the neighbors before you buy.

Short canals, a few tens of yards long, are generally much safer than long ones. Also, while most canals are initially deep enough for small craft traffic, sufficient sand movement on the back sides of barrier islands can result in the filling of the canals and subsequent navigation problems. On narrow islands, finger canals dug almost to the ocean side offer a path of least resistance to storm waters and are therefore potential locations for new inlets. Many such examples of future inlet or pass sites exist on Pinellas County islands. Finally, any homesite next to a finger canal is likely to be at very low elevation and should be evacuated immediately with the first storm warnings.

Following is a list of the characteristics that are essential to site safety:

1. Site elevation is above anticipated storm-surge level.
2. Site is behind a natural protective barrier such as a line of sand dunes.
3. Site is well away from a migrating inlet.
4. Site is in an area of shoreline growth (accretion) or low shoreline erosion. Evidence of an eroding shoreline includes: (*a*) sand bluff or dune scarp at back of beach; (*b*) stumps or peat exposed on beach; (*c*) slumped features such as trees, dunes, or man-made structures; (*d*) protective devices such as seawalls, groins, or pumped sand.
5. Site is located on a portion of the island backed by salt marsh.
6. Site is away from low, narrow portions of the island.
7. Site is in an area of no or low historic overwash.
8. Site is in a vegetated area that suggests stability.
9. Site drains water readily, even during winter.
10. Fresh groundwater supply is adequate and uncontaminated. There is proper spacing between water wells and septic systems.
11. Soil and elevation are suitable for efficient septic tank operation.
12. No compactable layers such as peat are present in soil below footings. (Site is not a buried salt marsh.)
13. Adjacent structures are adequately spaced and of sound construction.

County by county description of the beaches and barrier islands of Florida's Gulf Coast

Escambia and Santa Rosa counties (figures 5.13–5.15)

The 13.5 miles of Perdido Key (fig. 5.13) that lie in Florida contain the westernmost beaches in Florida. The reference map shows that with the exception of a 1.5-mile stretch about 3,000 feet wide, the 2,930-acre island is low and narrow. Beaches on the Gulf side vary in width from about 50 feet to 75 feet, and dunes are usually less than 10 feet high. Santa Rosa Island is east of Perdido Key, lying across most of the entrance to Pensacola Bay (figs. 5.14–5.16). Slightly more than 26 miles of this island lies in Escambia County and 3.1 miles in Santa Rosa County. Like Perdido Key, Santa Rosa Island is narrow over much of its length. Dunes range from heights of 6 to 7 feet on the ends to between 8 and 12 feet in the center. Sand bodies 30 to 35 feet high around Fort Pickens appear to be associated with old gun emplacements. The beaches of Santa Rosa Island are somewhat wider than those of Perdido Key, ranging up to about 125 feet. The foreshore and offshore profiles of both islands are relatively steep. Santa Rosa Island and Perdido Key had well-vegetated and anchored dunes before the overwash and flooding during Hurricane Frederic (1979) destroyed most of the dune grasses.

Extreme storm surge associated with hurricanes may raise the water levels 12 to 16 feet above normal along the open coast. The annual probability of a tropical storm striking these 2 counties is 21 percent; it is 13 percent for hurricanes. Some 15 to 20 frontal systems strike the coast every winter. Because of the long east-west fetch of Santa Rosa Island and Perdido Key, storms produce moderately high waves that refract into the shore. Resulting shoreline retreat has been about 2 feet per year for the last 100 years.

Hundred-year flood levels are estimated at 10.5 feet to 11.5 feet along the coast. More than a dozen major hurricanes have struck this coast in the last century. Some notable ones are the storms of 1896, 1901, 1906, 1911, 1916, 1917, 1926, 1932, 1947, 1950, 1969, and 1979. During the 1906 hurricane the water levels on Santa Rosa Island reached 11.6 feet above mean sea level at Fort Pickens, breached Santa Rosa Island just east of the Coast Guard Life Station, and claimed 151 lives in Florida and Alabama. The total damage exceeded $3.2 million—quite a sum at that time. The storm of 1926 caused 10.5-foot tides at Fort Pickens and more than $4.7 million in damages.

During Hurricane Frederic in 1979 storm surges of 14.0 feet and 12.0 feet above mean sea level were recorded at Perdido Key and Pensacola Beach, respectively. More than 90 percent of Santa Rosa Island and Perdido Key were under water. Eastern Santa Rosa Island suffered dune erosion and extensive washover. There was substantial damage to Fort Walton Beach pier and several dune walkways. But there was minor damage to many of the buildings due to good setbacks and distance from the center of the hurricane. Eglin Air Force Base property had 3 washover fans and minor to moderate dune erosion. At Navarre Beach the man-made channel was breached. Navarre pier suffered extensive damage to the outer 100 feet and 28 pilings. The outer pier deck was 18.0 feet above mean sea level and was damaged by wave uplift forces—indicating wave crest elevation of greater than 18 feet.

PROBLEMS
low
narrow in places
erosion potential: H=high
                    X=extreme

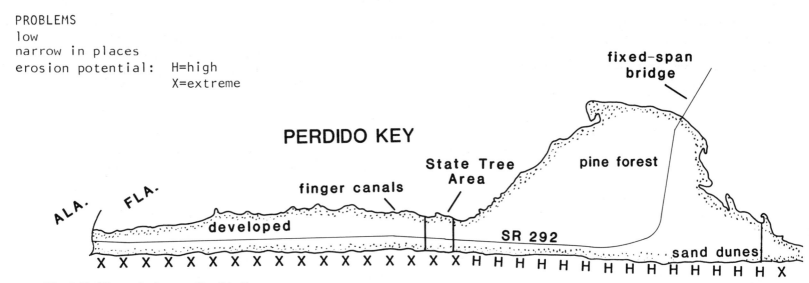

Fig. 5.13. Site analysis map: Perdido Key.

The Navarre Motel and the Holiday Inn suffered minor damage. At several locations the road was damaged due to flooding and washovers. More than 67 structures along the beach suffered some damage. Prudent observance of setback requirements and distance from the storm center kept property losses from being much greater. Erosion of dunes and beaches was minor to moderate.

During the 1979 storm the National Park Service area between Navarre pier and Pensacola Beach had 60 to 70 washover fans and experienced substantial damage to roads. Pensacola Beach pier was essentially destroyed. There was considerable damage to

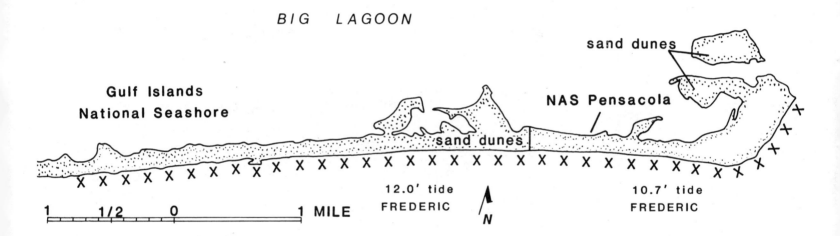

BIG LAGOON

Gulf Islands
National Seashore

sand dunes

NAS Pensacola

sand dunes

12.0' tide
FREDERIC

10.7' tide
FREDERIC

N

1   1/2   0   1 MILE

roads and utilities, and more than 128 structures fronting on the Gulf were damaged. There was moderate dune erosion, and large quantities of sand were transported across the island by flood-waters.

On the Fort Pickens Park section of shoreline extensive dune erosion and flooding occurred. Much of the road, utilities, and public facilities suffered severe damage. Several washover fans, 0.5 to 2 miles wide, were formed, and floodwaters carried sand from the beach to the bay.

The entire section of the national park on Perdido Key was

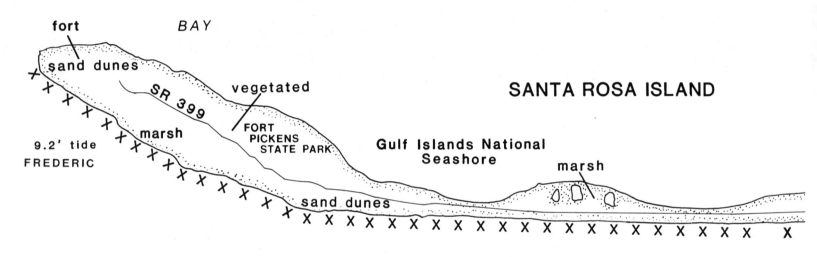

PENSACOLA BAY

fort

sand dunes

SR 399

vegetated

marsh

FORT PICKENS STATE PARK

9.2' tide
FREDERIC

sand dunes

SANTA ROSA ISLAND

Gulf Islands National Seashore

marsh

PROBLEMS
evacuation difficult   potential inlet: PI
low   erosion potential: H=high
rrow in places   X=extreme
tificially stabilized

Fig. 5.14. Site analysis map: western Santa Rosa Island.

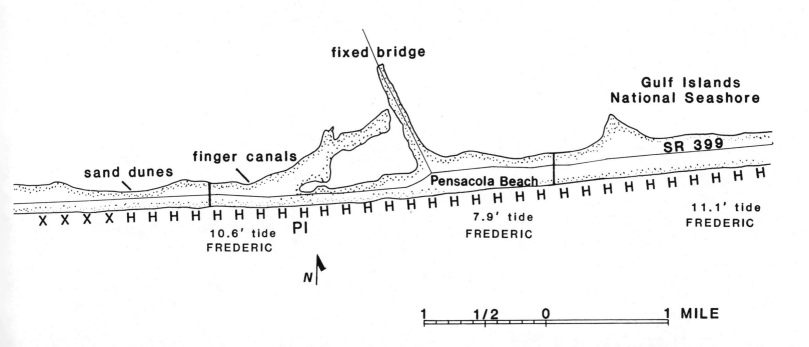

SANTA ROSA SOUND

fixed bridge

Gulf Islands
National Seashore

finger canals

SR 399

sand dunes

Pensacola Beach

X X X X H H H H H H H H H H H H H H H H H H H H H H H H H H H H

Pl

10.6' tide
FREDERIC

7.9' tide
FREDERIC

11.1' tide
FREDERIC

N

1   1/2   0   1 MILE

SANTA ROSA ISLAND

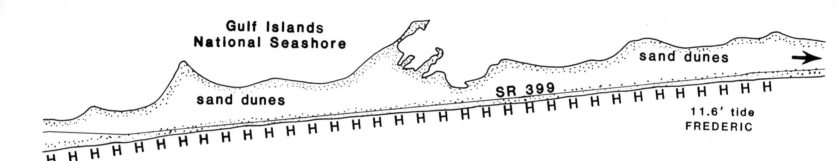

Gulf Islands National Seashore

sand dunes

sand dunes

SR 399

11.6' tide FREDERIC

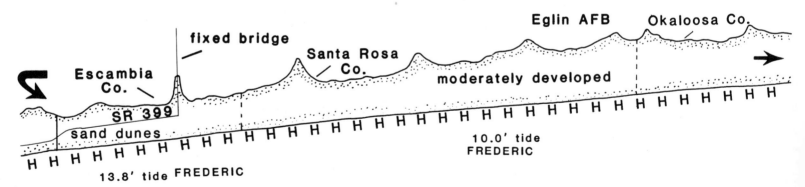

fixed bridge

Santa Rosa Co.

Escambia Co.

Eglin AFB

Okaloosa Co.

SR 399

sand dunes

moderately developed

13.8' tide FREDERIC

10.0' tide FREDERIC

Fig. 5.15. Site analysis map: central Santa Rosa Island.

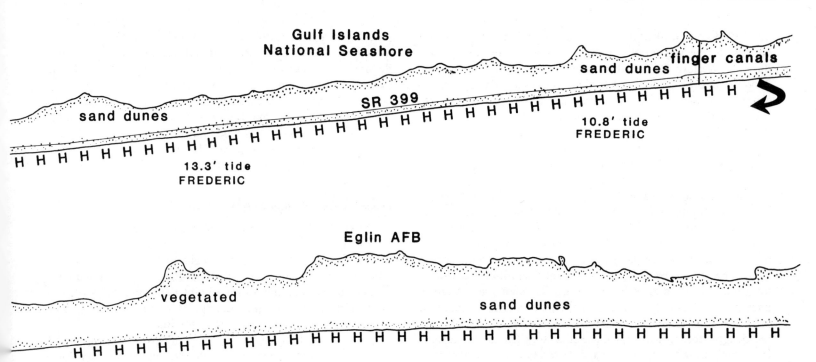

Gulf Islands
National Seashore

finger canals

sand dunes

sand dunes

SR 399

H H H H H H H H H H H H H H H H H H H H H H H H H H H H H H H H H H H H H H H H

10.8' tide
FREDERIC

13.3' tide
FREDERIC

Eglin AFB

vegetated

sand dunes

H H H H H H H H H H H H H H H H H H H H H H H H H H H H H H H H H H H H H H H H

1   1/2   0   1 MILE

flooded and overwashed in Hurricane Frederic, causing extensive damage to the vegetation. However, there was no inlet break-through. Western Perdido Key suffered extensive dune erosion, flooding, and property damage. The total dollar damage in Escambia County from Frederic exceeded $62 million, and for Santa Rosa County it exceeded $9.4 million. It is necessary to remember that Hurricane Frederic had a landfall 50 miles west of Pensacola, and a direct hit would have caused much greater damage. As of August 1981, 6,106 flood insurance policies valued at $418 million were in effect in the 2 counties.

The shorelines of Santa Rosa Island and Perdido Key have a history of buildup and erosion, with erosion predominating. Between 1867 and 1935, with the exception of the 2.6-mile west end of the beach near Pensacola Pass, the eastern half of the beach receded 100 to 200 feet landward, and the western half receded 3.0 feet per year. The western 2.6 miles at Fort Pickens advanced toward the Gulf and 2,500 feet westward. The shoreline appears to have retreated slightly during the 1970s, but the exact magnitude of this change is unavailable. From 1876 to 1963 Perdido Key receded 500 feet westward, while Santa Rosa Island was advancing, which caused undermining and destruction of the brick structure of Fort McRae. For the next 5 miles during the period from 1860 to 1935 the Gulf shoreline retreated 300 to 500 feet landward.

The Gulf beaches along Santa Rosa Island and Perdido Key have undergone little man-made interference—except for the navigational jetties at East Pass and the artificially deepened (35 feet), 500-foot wide Pensacola Bay channel. Navigational channels have compounded and increased the natural retreat rates west of the passes. However, because the properties west of these passes are owned by the federal government and managed as natural areas, the problems of erosion have not been as serious in financial terms as they would have been had these sections been developed.

Erosion problems are considered more serious along Santa Rosa Sound, Escambia Bay, and Big Lagoon behind the barrier islands. Much of this erosion is attributed to the drowning of the bay shoreline due to the rise in sea level. The Army Corps of Engineers estimates an erosion rate of about 2 feet each year along Santa Rosa Sound.

Okaloosa County (figures 5.15-5.17)

The Gulf of Mexico shoreline of Okaloosa County has 25 miles of beaches: about 18 miles of barrier beach on Santa Rosa Island west of the East Pass navigational channel (figs. 5.15 and 5.16), 2 miles of barrier spit east of the East Pass channel, and about 5 miles of beaches on the Moreno Point Peninsula (fig. 5.17). The Santa Rosa Sound–Choctawhatchee Bay shoreline has 45 miles of beaches and 43 miles of marshy shoreline.

The beaches along the Gulf are generally wide and flat, backed by foredunes 6 to 12 feet high with even higher primary dunes. For a distance of 2 miles on either side of the East Pass channel the beaches lack foredunes or primary dunes. The shoreline of Choctawhatchee Bay in Okaloosa County is primarily sandy beach with a low berm somewhat narrower than that of the Gulf beaches.

The chance of a hurricane striking this county is about 14 percent in any given year. The most severe hurricanes were in 1887, 1889, 1896, 1906, 1926, 1936, 1948, 1953, 1956, and 1975. The

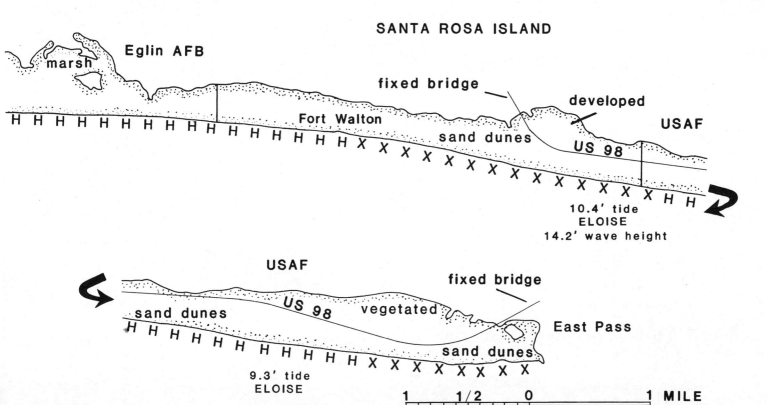

Fig. 5.16. Site analysis map: eastern Santa Rosa Island.

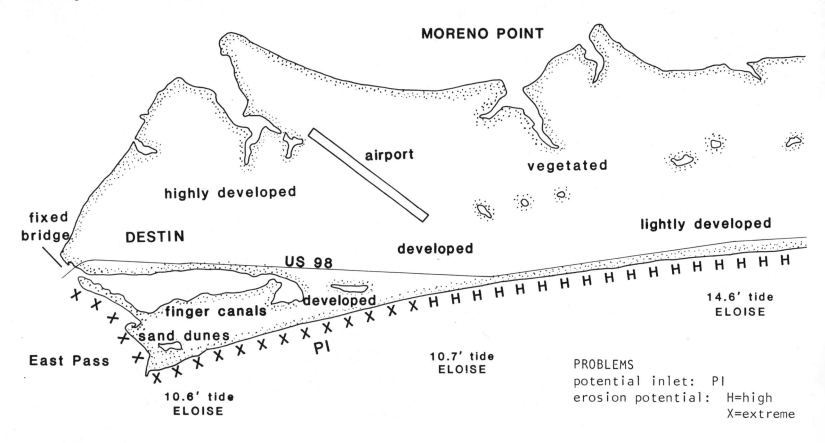

Fig. 5.17. Site analysis map: Moreno Point.

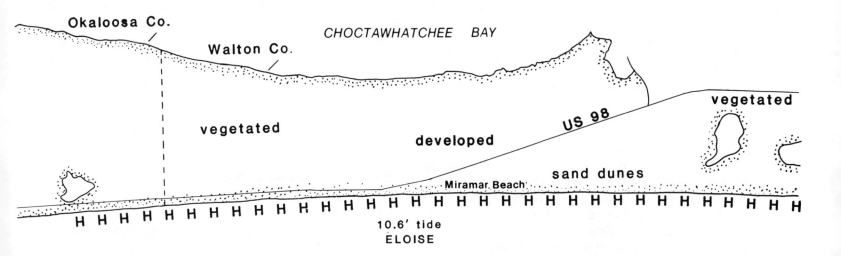

Okaloosa Co.

Walton Co.

CHOCTAWHATCHEE BAY

vegetated

developed

US 98

vegetated

sand dunes

Miramar Beach

vegetated

10.6' tide
ELOISE

H H H H H H H H H H H H H H H H H H H H H H H H H H H H H H H H H

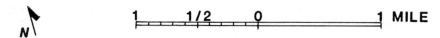

N

1   1/2   0   1 MILE

hundred-year flood levels are estimated at 9 to 10 feet above mean sea level, although during Hurricane Eloise in 1975 the barrier beaches experienced high tides of 9.5 to 14.4 feet above mean sea level along the Gulf and about 6 to 7 feet in Choctawhatchee Bay and Santa Rosa Sound. There was extensive damage to coastal properties along the Gulf and bay beaches due to flooding, with total damage exceeding $10 million.

The shoreline along the Gulf as well as in the bays has been retreating. East Pass was formerly about 2 miles east of its present location, and Santa Rosa Island's eastern tip extended to Norreigo Point across from Destin. Between 1871 and 1929 the seaward end of the pass migrated about 2,500 feet westward, and the original pass shoaled and closed sometime between 1935 and 1938. The present East Pass was formed in 1928, when a severe storm with high tides breached the narrow, low portion of Santa Rosa Island. Comparison of the shoreline in 1934–35 and 1962 shows that for a distance of 10 miles in either direction of the present East Pass, buildup averaged about 200 feet except for a 1.5-mile section near the pass where erosion was predominant. The maximum range for buildup or erosion was about 300 feet. The western 8 miles of Santa Rosa Island shoreline retreated landward 150 to 300 feet between 1872 and 1927 and advanced toward the Gulf 100 to 150 feet from 1927 to 1935, with a net retreat of 50 to 150 feet from 1872 to 1935. The 2-mile spit east of the pass (across from Destin) is very unstable and low but has built up since the 1930s.

During the last decade development on Okaloosa County beaches has greatly intensified. Many buildings have been placed in the areas that overwashed and flooded repeatedly during this century. Hurricane Eloise (1975) flooded and overwashed the area around East Pass, especially on the barrier spit opposite Destin, causing extensive scouring and erosion. More than 2,823 federally subsidized flood insurance policies valued at more than $229 million were in effect as of August 1981. A major hurricane is likely to inflict heavy economic losses to buildings. The construction of permanent buildings on top of old inlets and washover sites is not prudent but has been permitted in Okaloosa County.

Walton County (figures 5.17–5.18)

Walton County has more than 22 miles of shoreline along the Gulf of Mexico and 55 miles of beach and 11 miles of marsh shoreline on Choctawhatchee Bay. Many of the county's beaches are mainland beaches. The eastern 3 miles have 12- to 18-foot dunes, the central 12 miles have 20- to 30-foot dunes, and the western 7 miles have 12- to 20-foot dunes. In general, the terrain north of the beaches tends to be made up of marsh and swamps with many small streams that result in a series of small to medium freshwater lakes along the coast. Intermittently, openings through the dune fields connect some of these lakes to the Gulf of Mexico.

Detailed information on winds, tides, waves, currents, and storms has not been compiled for this coast due to sparse population and development. However, general characteristics are similar to those of Bay and Escambia counties.

The probability of a hurricane striking the Walton County coast is 7 percent per year, and 100-year flood levels have been estimated at 8.5 feet above mean sea level. However, during Hurricane Eloise in 1975 a storm surge of 11 to 18 feet was recorded. Hurricane Eloise caused extensive scouring and erosion of dunes and beaches, but the total damage to coastal property and structures was only a relatively low $3 million—primarily because of scanty development and an observance of the coastal construction setback line.

During Hurricane Eloise several bridge approaches over the coastal creeks and streams emptying into the Gulf were washed away by floodwaters. Grayton Beach State Park beaches and dunes suffered severe scouring and erosion by 16-foot waves. During Hurricane Frederic in 1979 the entire Walton County beach front on the Gulf experienced loss of foredunes and erosion of primary dune lines to about 10 feet horizontally and 3 feet vertically. There were several washovers at the discharge outlets of coastal lakes and lagoons. As urban development proceeds on this coast, the losses to structures and property will increase, even with good setbacks, because the shoreline is naturally retreating.

From 1872 to 1935 the Gulf shoreline of Walton County retreated by 100 to 299 feet. In recent years the shoreline has experienced an average of 1 to 2 feet of erosion per year, similar to that of western Bay County. Erosion around Choctawhatchee Bay, which has more development than the beach, is mainly due to the drowning of the bay shoreline by the rising sea level. Bay erosion is considered most serious in the communities of Villa Tasso and Choctow Beach. On the Gulf coast serious shoreline recession has been experienced in the areas around Inlet Beach, Seagrove Beach, Grayton Beach State Park, Dune Allen Beach, and Miramar Beach.

As of August 1981 there were 1,155 federally subsidized flood insurance policies valued at $73,556,000 covering portions of Walton County's coast.

Bay County (figures 5.18–5.21)

Bay County's Gulf of Mexico shoreline is characterized by Shell Island (fig. 5.20) with 11 miles of beachfront, 22 miles of mainland shoreline, and 3,100 acres of land; Dog Island with 2 miles of beaches and about 150 acres; and Crooked Island (fig. 5.21), 9.5 miles long with 1,700 acres.

The barrier beaches are composed of fine quartz sand and shell fragments in varying proportions. Almost all of the mainland and island beaches are backed by dunes that in some cases can be as high as 25 feet and at other times lower than 5 feet. Shell Island is the widest of the barrier islands and has 3 distinct sections. The oldest section is made up of wide dune ridges separated by wide swales containing peat deposits. The section of intermediate age is composed of narrow beach ridges separated by narrow swales that connect the barrier to the mainland. Both sections have elevations of 8 to 12 feet. The youngest deposits comprise the eastern peninsular area, a low beach ridge and narrow sand flat with 5- to 10-foot elevations. Crooked Island, most of which has emerged since 1779, is composed of low beach ridges 5 to 7 feet high at the outward building northwestern end and at the old inlet sites, whereas

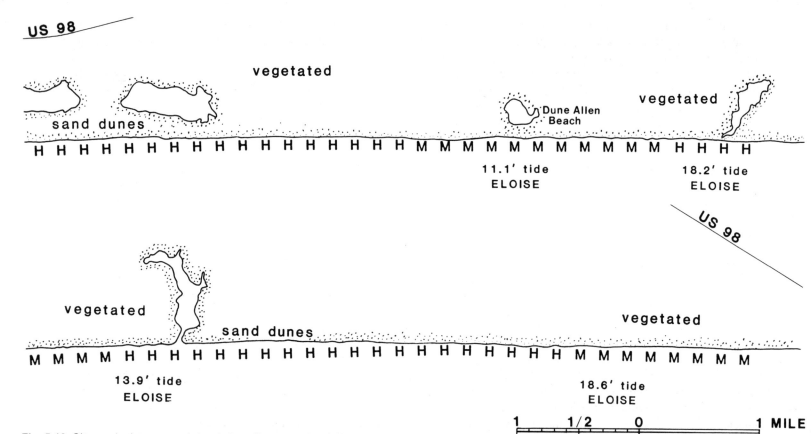

Fig. 5.18. Site analysis map: mainland shoreline near the Walton-Bay County line.

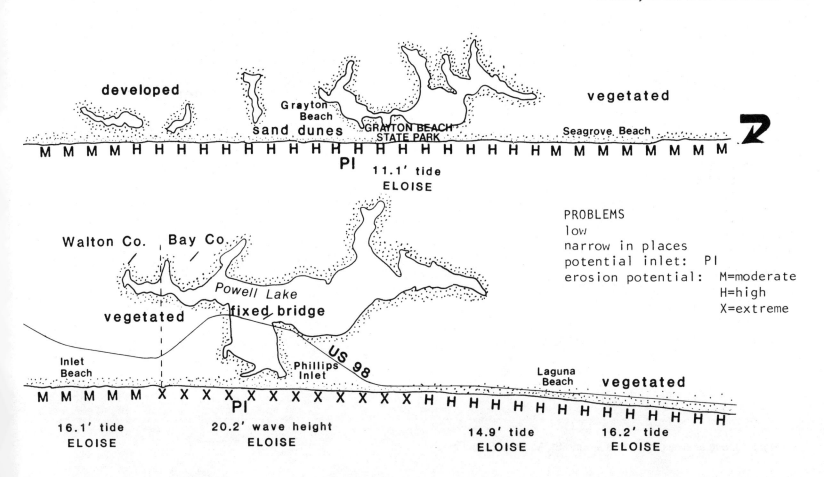

developed

Grayton
Beach

sand dunes

GRAYTON BEACH
STATE PARK

vegetated

Seagrove Beach

M M M M H H H H H H H H H H H H H H H H H M M M M M M M

PI

11.1' tide
ELOISE

Walton Co.   Bay Co.

Powell Lake

vegetated   fixed bridge

US 98

Inlet
Beach

Phillips
Inlet

Laguna
Beach

vegetated

M M M M M X X X X X X X X X X X X X H H H H H H H H H H H H H

PI

16.1' tide
ELOISE

20.2' wave height
ELOISE

14.9' tide
ELOISE

16.2' tide
ELOISE

PROBLEMS
low
narrow in places
potential inlet: PI
erosion potential: M=moderate
                    H=high
                    X=extreme

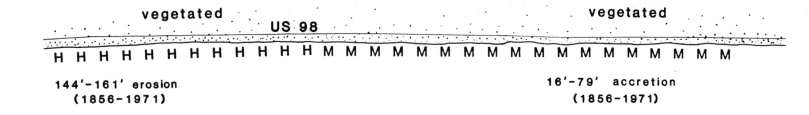

vegetated   US 98   vegetated

H H H H H H H H H H H H H M M M M M M M M M M M M M M M M M M M M M M M

144'-161' erosion   16'-79' accretion
(1856-1971)   (1856-1971)

PROBLEMS
low
artificially stabilized
potential inlet: PI
erosion potential: M=moderate
                   H=high
                   X=extreme

Fig. 5.19. Site analysis map: area around St. Andrews State Park.

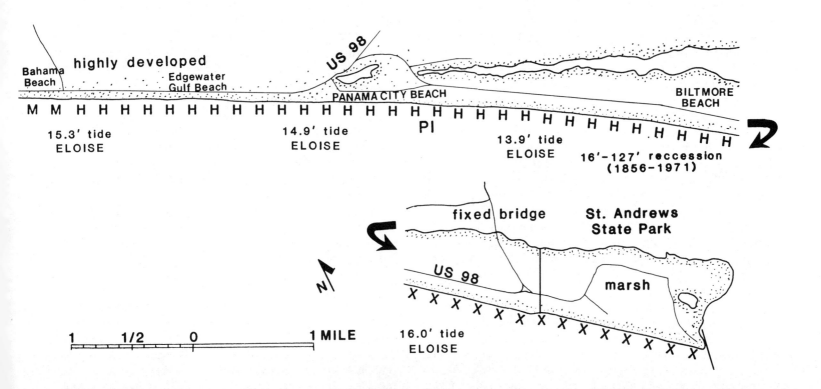

highly developed

Bahama
Beach

Edgewater
Gulf Beach

US 98

PANAMA CITY BEACH

BILTMORE
BEACH

M   M   H   H   H   H   H   H   H   H   H   H   H   H   H   H   H   H   H   H   H   H   H   H   H   H   H   H   H   H

15.3' tide
ELOISE

14.9' tide
ELOISE

PI

13.9' tide
ELOISE

16'-127' reccession
(1856-1971)

fixed bridge

St. Andrews
State Park

N

US 98

marsh

1   1/2   0   1 MILE

X   X   X   X   X   X   X   X   X   X   X   X   X   X   X   X

16.0' tide
ELOISE

high dunes (12 to 25 feet) predominate along the retreating narrow section (less than 550 feet wide). Several high dune areas now mark the position of old filled passes/inlets. Most of the natural beaches are 100 to 200 feet wide.

The major changes in the area's barrier beaches take place during severe hurricanes and storms, but slow and continuing changes occur as a result of wind-generated waves, longshore currents, rise in sea level, and emplacement of man-made structures for erosion "control" and navigational purposes.

Extreme storm surge associated with hurricanes may raise the water levels on this shoreline 10 to 20 feet above mean sea level. From 1899 to 1921, 21 major hurricanes and more than 82 tropical storms and smaller hurricanes have hit the Panama City Beach area, or an average of about once every 4 years for big hurricanes and once every 1.5 years for tropical storms and small hurricanes.

Hurricane Eloise, which struck Bay County on September 12, 1975, was one of the most severe in northwest Florida's history. The strongest wind gusts near Panama City exceeded 155 mph, with sustained winds of 130 mph. Barometric pressure fell to 28.2 inches. During the storm, the tides ranged from 12 to 16 feet above normal from just east of Fort Walton Beach (Okaloosa County) to Mexico Beach east of Panama City Beach. The highest water mark surveyed by the U.S. Army Corps of Engineers exceeded 18.6 feet of mean sea level 3 miles west of Phillip's Inlet. Waves intensified by winds and high tides reached a maximum of 20.2 feet above mean sea level, with numerous occurrences in excess of 18.0 feet where Eloise made landfall along the coastal areas. Rainfall averaged 5.5 inches over the course of the storm.

Other notable storms that hit Bay County include one of August 30, 1856, a storm with recorded tides of 10 feet above mean sea level. During this storm West Pass (on Shell Island) opened, then was closed again in 1861 and remained so until 1881. A storm on September 27, 1906, produced tides 6 to 10 feet above normal in Panama City, causing some property damage. Other major hurricanes to hit the coastline occurred in 1924, 1926, 1929, 1936, 1950, 1953 (Florence), 1956 (Flossy), and 1972 (Agnes). Unusual weather conditions from October 5–11, 1970, caused severe erosion of beaches throughout the entire county. On the average, 15 to 20 frontal systems during winter months produce serious beach retreat and scouring.

Shoreline surveys for all of Bay County for 1855, 1934, 1945, and 1968–70 provide the basis for the following summary of the history of advance and retreat of the shoreline. The historic data show that the shoreline from the Bay County–Gulf County line to the base of Crooked Island advanced toward the Gulf by between 200 to 300 feet from 1855 to 1934. From 1935 to the present the eastern shoreline retreated, but the amount of this recession is not clear. At the northwest end of Mexico Beach (at the base of Crooked Island; fig. 5.21), a small boat channel was dug through the dunes in 1956, which was reinforced with a wooden jetty in 1960 and with a concrete rubble extension of 60 to 70 feet in 1961–62. This littoral barrier has created some erosion problem on the eastern side of the channel on Mexico Beach. The rest of eastern Bay County is more stable, with higher erosion rates on

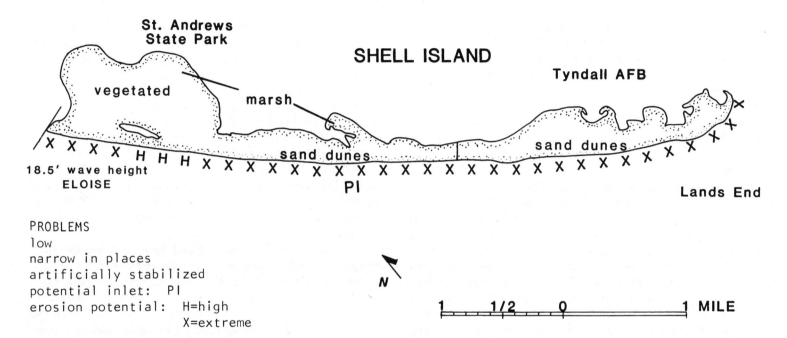

St. Andrews
State Park

SHELL ISLAND

vegetated

Tyndall AFB

marsh

X X X X X H H H X X X X X X X X X X X X X X X X X X X X X X X X X X X X X X X X X X

sand dunes

sand dunes

18.5′ wave height
ELOISE

PI

Lands End

PROBLEMS
low
narrow in places
artificially stabilized
potential inlet: PI
erosion potential: H=high
                    X=extreme

N

1 ____ 1/2 ____ 0 _____ 1 MILE

Fig. 5.20. Site analysis map: Shell Island.

the urbanized coast due to man's structures and activities.

Except for the western end, Crooked Island has a history of buildup and retreat, with buildup predominating. Between 1779 and 1970 Crooked Island migrated in easterly as well as westerly directions with frequent opening and closing of passes. Between 1855 and 1934 the western end of Crooked Island retreated 9,300 feet, and the outer shoal migrated more than 3,700 feet landward. The mainland beach behind St. Andrew Sound also built up by 1,100 feet. The entrance to St. Andrew Sound was reduced from about 2,100 feet in width to only 500 feet by 1934. The eastern end of Crooked Island was connected to St. Andrew Point (or Cape False) between 1779 and 1829.

The most impressive change in the area is the appearance of Dog Island in 1954–55 from a shallow shoal—without a hurricane occurring. This island was attached to the mainland during the 1968–70 period.

Shell Island was an 11-mile-long barrier spit extending east from Panama City Beach and enclosing St. Andrew Bay. In 1934 the federal government opened a deep-draft channel 34 feet deep and 450 feet wide in about the middle of Shell Island. The east and west rubble mound jetties extend 2,075 feet and 2,700 feet, respectively. The west jetty was breached during Hurricane Eloise in 1975, which generated 18.5-foot waves. The federal share of the cost for the construction of this jetty was $1.638 million, and operation and maintenance costs as of 1980 totaled $3.84 million.

The eastern section of Shell Island has undergone many rapid changes—buildup, erosion, breaching, and migration. The shoreline migrated landward 2,000 feet between 1855 and 1934. The 1855 survey showed a 1,600-foot-wide inlet about 2 miles west of the present eastern tip of the island. By 1870 this pass was closed, and the Gulf shoreline had moved landward by 400 feet. The 1902 survey indicated another wide inlet about 1.5 miles east of the 1855 inlet; that barrier spit extended 6,000 feet east of the present tip of the island. Shell Island has rotated landward around an axis near the present jetties; the eastern half of the island has moved landward 2,100 feet, building out with a crescent-shaped beach on the east end.

The entire 18.4 miles of Bay County beaches west of the jetties have experienced gradual retreat since records have been kept. According to the Army Corps of Engineers, from St. Andrew State Park to Oriole Street (2.5 miles west; fig. 5.19) the beaches retreated 187 to 97 feet during the period from 1856 to 1971, with an overall erosion rate of 0.7 feet per year for the whole stretch. From Biltmore Beach west to a point east of Bahama Beach, shoreline retreat has ranged from 127 feet to 16 feet during the 1856–1971 period, with an overall erosion rate of 1.5 to 2.5 feet per year. At the same time a 3-mile section from Bahama Beach to Gulf Resort Beach advanced 16 to 79 feet. The western 6.5 miles to Phillip's Inlet near the borderline between Bay and Walton counties experienced retreat of 51 to 251 feet, most severely along the 2 miles on either side of Phillip's Inlet.

Hurricane Eloise's storm surge and waves undercut the dunes and scoured the beaches from Mexico Beach to Phillip's Inlet. The extent of vertical loss shown in beach profiles ranged from a low of 2 feet to as high as 8 feet, and horizontal migration ranged from a minimum of 20 feet to a maximum of 150 feet. The scouring of the beaches and dune profiles undermined the foundation of many buildings—especially slab-on-grade structures—causing great loss of private and public coastal property. In Bay County losses exceeded $88 million. Since 1975 Bay County beaches have been urbanized at a much greater rate, and as of August 1981 more than 4,000 federally subsidized flood insurance policies valued at $223 million were in effect countywide. A hurricane similar to Eloise would cause much greater losses to beaches and property now because of increased development and destruction of the protective dunes.

After Hurricane Eloise emergency measures were taken to dredge and open the inlet near Mexico Beach. Approximately 238,000 cubic yards of sand were placed on a 4.7-mile section of the beach from the entrance of Panama City Harbor to Phillip's Inlet. In 1979 Hurricane Frederic also caused dune retreat of about 10 to 15 feet horizontally and up to 2 feet vertically at Mexico Beach, while Panama City beaches lost an immense amount of sand, with a maximum erosion of 50 feet horizontally and 7 feet vertically. All of this happened with Frederic having a landfall *160 miles west* of Panama City Beach!

Construction of jetties on Shell Island for navigation purposes has contributed to some erosion problems west of the channel. The U.S. Army Corps of Engineers estimates that since the jetties were opened in 1934, more than 11.4 feet per year of retreat occurred for 2.6 miles west of the jetties. Between 1934 and 1945, immediately west of the jetty the rate was 19 feet per year. Since then the erosion rate has slowed down somewhat due to the placement of dredged materials from channel maintenance west of the jetties. However, the average erosion rate was still 7 feet per year for the period from 1935 to 1971. The jetties have changed the configuration of the offshore shoals and act as a sink for the sand, thereby depriving the western beaches of the sand necessary to replenish themselves. It has been estimated that more than 5 million cubic yards of sand have been taken out of the littoral sand-sharing system between 1935 and 1971. The effect of jetties diminishes as one proceeds farther west, and long-term beach retreat is probably due to sea-level rise.

It should be pointed out that flood and erosion damage to the beach dune systems and structures in Bay County during Hurricane Eloise were greater than in either Walton or Okaloosa counties to the west. This is attributed to less development in the latter 2 counties, and, more important, the losses in Walton and Okaloosa counties were minimized because of a better adherence to a sound coastal construction setback line that protected primary and secondary dunes. Bay County's coastal construction setback line chopped the dune crests in half and in many cases permitted leveling of the dunes to build bulkheads close to the water. Although Walton and Okaloosa counties suffered severe scouring of beaches and dunes, they were able to recover much more quickly with minimal losses to structures from floodwaters. The Army Corps of Engineers has recommended a beach nourishment program for the entire 18.5 miles of beach west of the St. Andrew Park jetty.

Gulf County (figure 5.21)

The beaches of Gulf County are characterized by barrier spits at Indian Peninsula and St. Joseph Spit and by mainland beaches. The 3-mile section from the eastern tip of Indian Peninsula to its base at Camp Palm has beaches 90 to 100 feet wide with low dunes. Mainland beaches from Camp Palm to the base of St. Joseph Spit are about 200 feet wide and backed with 10- to 30-foot dunes. For about 3 miles to the west from the base of St. Joseph Spit to Cape San Blas, the headland beach is backed by multiple

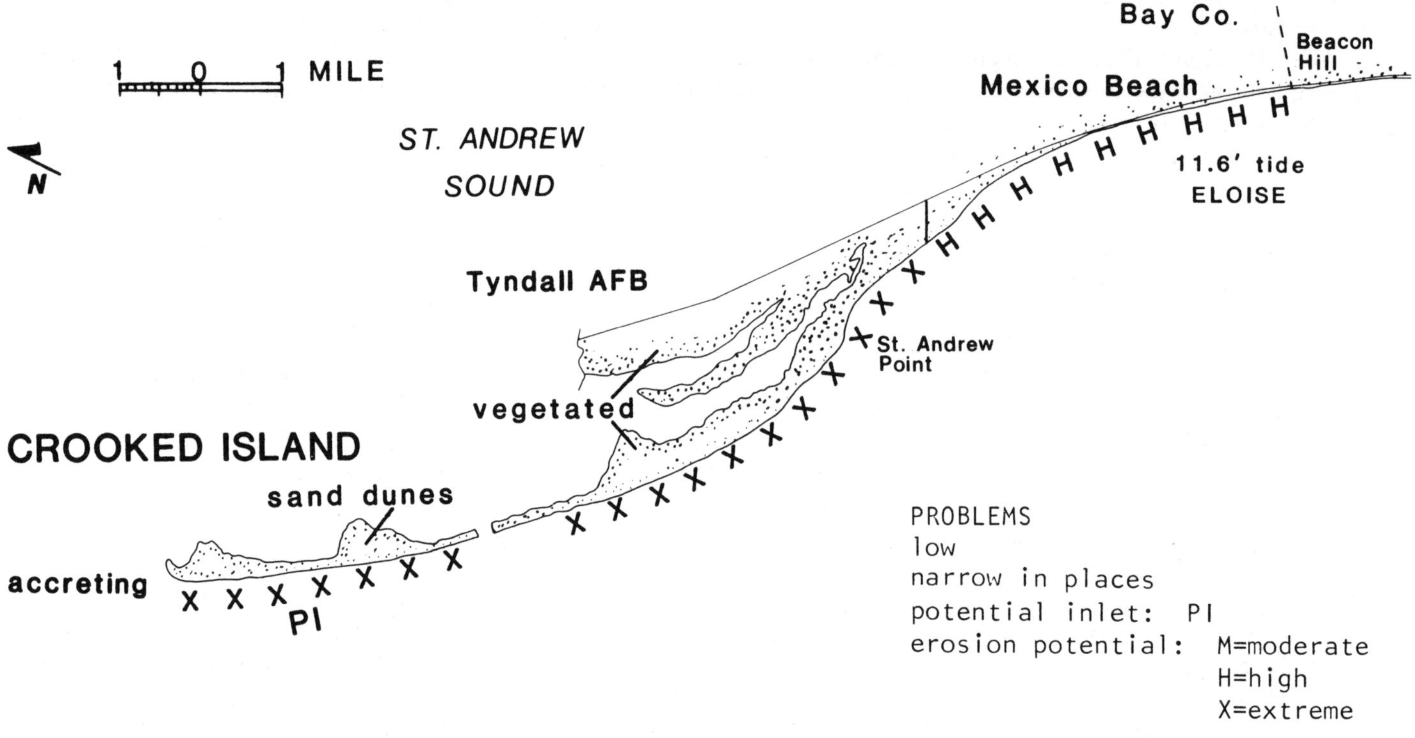

Fig. 5.21. Site analysis map: Crooked Island through Indian Peninsula.

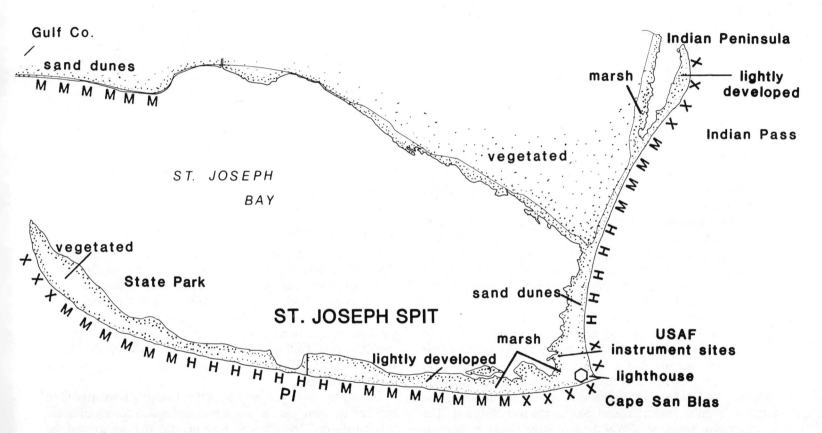

Gulf Co.

sand dunes
M M M M M M

ST. JOSEPH
BAY

vegetated

State Park

vegetated

X
X
X M
X
M
M
M
M
M
H
H
H
H
H
H
H
Pl
H
H
M
M
M
M
M
M
M
X
X
X
X

ST. JOSEPH SPIT

lightly developed

sand dunes

marsh

vegetated

marsh

Indian Peninsula

lightly
developed

Indian Pass

M
M
M
M
M
H
H
H
H
H
H
H
X
X

USAF
instrument sites

lighthouse

Cape San Blas

dunes 10 to 25 feet in elevation. At Cape San Blas the spit turns sharply and extends for about 18 miles in a northerly direction. The southern 3 miles of St. Joseph Spit is backed by 10-to 15-foot dunes, and the remaining northern section has dunes 20 to 30 feet high. In 1 spot dunes in excess of 50 feet are found. Most of St. Joseph Spit is about 0.5 miles wide, with the narrowest section being about 800 feet wide at Eagle Harbor. A hurricane in 1844 breached the spit and destroyed the original town of Port St. Joe. More than 12 hurricanes have passed over Gulf County between 1886 and 1975, with a hurricane frequency of once every 7 years. The 100-year flood levels are estimated at 9.5 to 12.5 feet above mean sea level.

Available information on the shoreline changes for Gulf County is limited. The shoreline of the entire county is apparently undergoing landward retreat. The shoreline of Indian Peninsula has retreated 100 feet from 1855 to 1945. Westward toward Cape San Blas to the base of the spit the shoreline appears to have advanced. Generally, the western 3 miles to the Cape have eroded, but there has been intermittent localized buildup.

Cape San Blas is somewhat peculiar in that while there has been erosion on the eastern as well as the northern beach, the cape itself has built out southward into the gulf. St. Joseph Spit north of Cape San Blas has undergone erosion and moved eastward more than 2,300 feet in the immediate vicinity of the cape. The adjacent 4 miles have retreated 200 to 300 feet eastward. The southernmost 3-mile portion of the cape is the most rapidly eroding shoreline in Florida—averaging 25 feet per year for the 150-year period when records have been kept. The lighthouse has been moved 3 times in the last 100 years because of this severe erosion. Landward retreat for the entire spit averaged about 100 to 200 feet during the last century, except for the northern 2.5-mile section terminating in St. Joseph Point, which has advanced westward and northward. The beaches on the mainland east of St. Joseph Spit and to the north advanced about 200 to 300 feet for the period from 1855 to 1935.

Much of Cape San Blas and St. Joseph Spit is in public ownership. However, if development in the privately owned section proceeds, it will be necessary to recognize the high natural erosion rates.

Franklin County (figures 5.22–5.26)

The barrier beach pattern is maintained in Franklin County, where from west to east we find St. Vincent Island (fig. 5.22), St. George Island (figs. 5.23 and 5.24), Dog Island (fig. 5.25), and the St. James Island–Lighthouse Point–Bald Point complex (fig. 5.26). The shoreline then changes dramatically to the east and south in the Big Bend of Florida where it is characterized by a transition from Gulf to land dominated by small islands covered with marsh and mangrove, some with limestone outcroppings, and by small beaches.

During the last 130 years Franklin County's coast has been battered by more than a dozen hurricanes and dozens of lesser tropical storms. These storms have flooded and overwashed the islands with storm surges 9 to 15 feet above normal. Hundred-

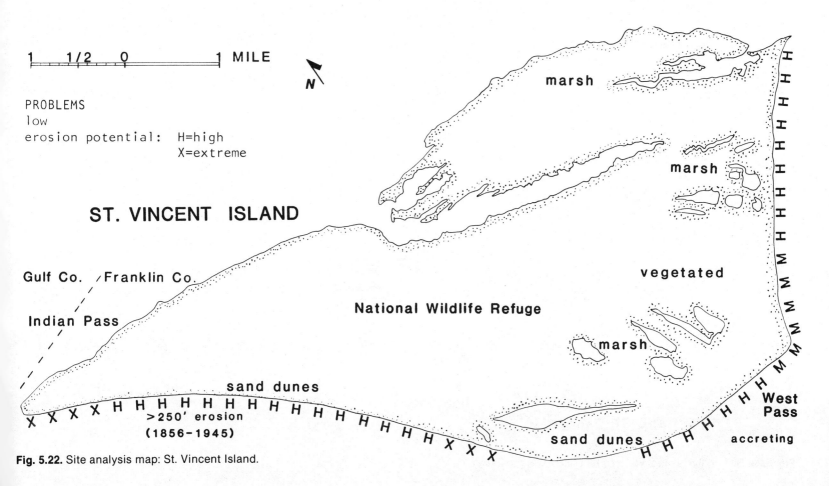

1   1/2   0   1 MILE

N

PROBLEMS
low
erosion potential: H=high
                     X=extreme

ST. VINCENT ISLAND

marsh

marsh

vegetated

Gulf Co. Franklin Co.

National Wildlife Refuge

Indian Pass

marsh

HHHHHHHHHHHHHHHHHMMMMMMMMMMM

West
Pass

sand dunes

sand dunes

accreting

X X X X X H H H H H H H H H H H H H H H H H H X X X
>250' erosion
(1856-1945)

Fig. 5.22. Site analysis map: St. Vincent Island.

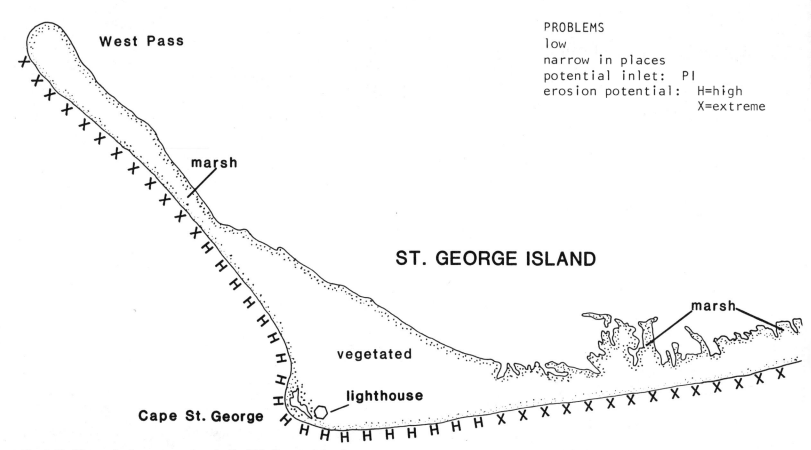

West Pass

PROBLEMS
low
narrow in places
potential inlet: PI
erosion potential: H=high
                    X=extreme

marsh

ST. GEORGE ISLAND

marsh

vegetated

lighthouse

Cape St. George

Fig. 5.23. Site analysis map: western half of St. George Island.

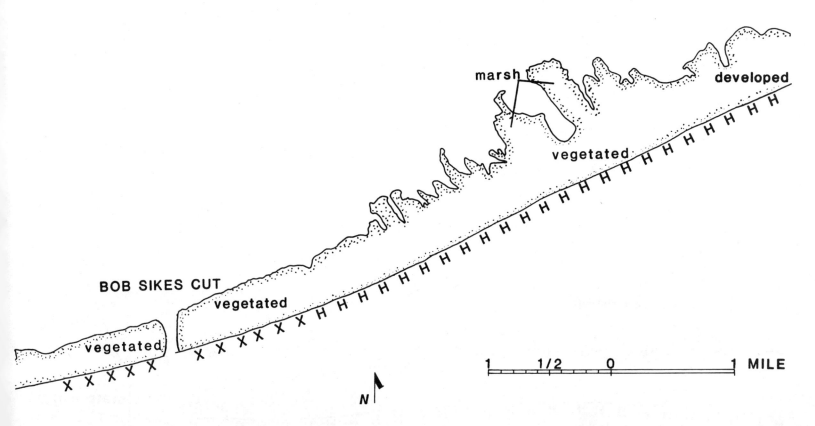

BOB SIKES CUT

marsh

vegetated

developed

vegetated

vegetated

1   1/2   0   1   MILE

N

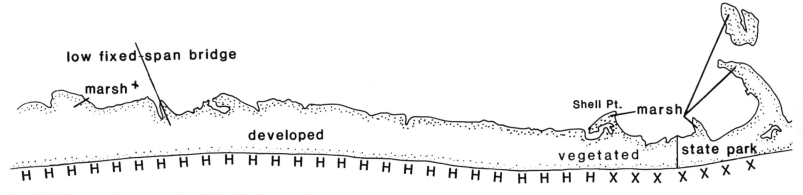

ST. GEORGE ISLAND

low fixed-span bridge

marsh +

developed

Shell Pt. marsh

vegetated

state park

H H H H H H H H H H H H H H H H H H H H H H H H H H H H H H X X X X X X X

Fig. 5.24. Site analysis map: eastern half of St. George Island.

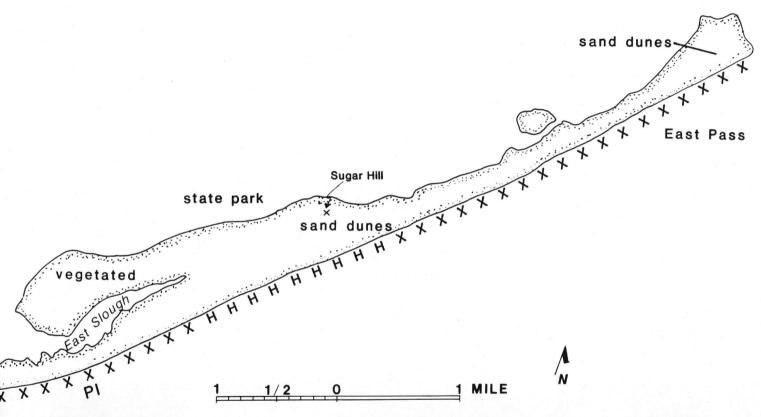

sand dunes

East Pass

Sugar Hill

state park

sand dunes

vegetated

East Slough

X X X X X Pl

1   1/2   0   1 MILE

N

300'-400' erosion
during previous 70-80 yrs.

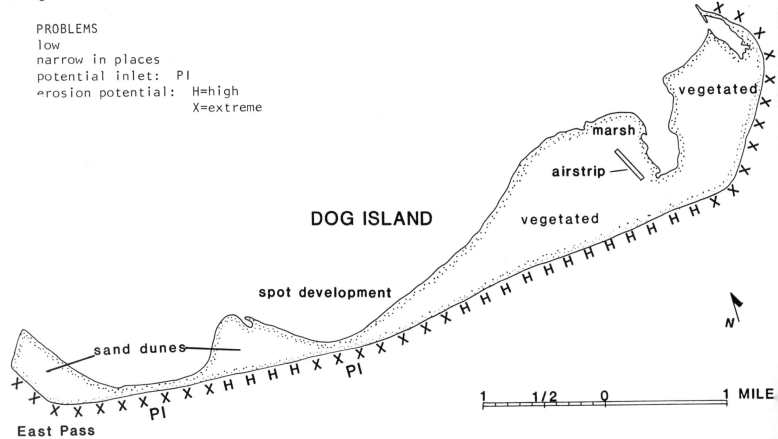

PROBLEMS
low
narrow in places
potential inlet: PI
erosion potential: H=high
X=extreme

DOG ISLAND

vegetated

marsh

airstrip —

vegetated

spot development

sand dunes —

PI

PI

East Pass

1   1/2   0   1 MILE

N

Fig. 5.25. Site analysis map: Dog Island.

year flood levels are estimated at 12 to 14 feet along this county's coast.

Much of the Alligator Harbor Peninsula is fronted with 8- to 10-foot dunes. Beaches in this area have been relatively stable and average about 50 feet in width. The area from Bald Point to midway between Bald Point and Lighthouse Point has advanced eastward, while the southern half has retreated westward. The magnitude of this advance and retreat has been 200 to 300 feet between 1856 and 1943. The rest of the barrier spit has retreated landward and advanced westward. There are several low-lying areas that are subject to overwash during storms. Recent man-made canals and land clearings have produced 3 or 4 potential inlet sites for a hurricane to exploit.

Dog Island (fig. 5.25) has undergone constant landward migration of the shoreline for all except the east and west ends where it is building outward. The beaches are rather narrow, and the offshore Gulf profile is moderately steep. Three-quarters of the island is protected by dunes averaging 8 to 23 feet in height. Dog Island is an excellent example of curving dune-and-swale topography in a relatively undisturbed state. The highest dunes are more than 35 feet. The western areas of the island have been subject to frequent overwash during minor storms because of their low topography. A 3-mile section fronting the Gulf that has been developed with homes is undergoing erosion, and steep scarps are appearing on the dunes. The west end of Dog Island has advanced about 1 mile between 1855 and 1935.

St. George Island's (figs. 5.23 and 5.24) eastern end built outward more than 1 mile between 1855 and 1935. At the same time the beach from the eastern tip to about Sugar Hill has retreated 300 to 400 feet landward. This is evident from scarps in the frontal dunes. The next small section of about 2 miles has wider dunes and may be somewhat more stable. The area between East Slough and Shell Point has a history of erosion, frequent overwash, and breaching during storms. The remaining section of the island to about 1.5 miles east of Sikes Cut (Sikes Cut was opened in 1964) has advanced toward the Gulf on the average of about 100 feet from 1855 to 1935. For the next 3 miles to the old inlet, which was surveyed in 1856–57 and closed sometime before 1935, the shoreline has retreated. From the old inlet site of Cape St. George Lighthouse, the mean high water shoreline also migrated landward. Cape St. George built out in a southwesterly direction 1,600 feet during the period from 1856 to 1943, but recently it also has begun to retreat landward. The mean high water shoreline of the northerly extension generally advanced seaward from 1856 to 1935. A breach occurred in the middle of this section in 1855, and an inlet was opened that was closed due to longshore drift by 1902. The ends of the island and the low points near the old inlet sites, as well as the section between East Slough and Shell Point, are flooded by 10-year storm tides. Large sections of this island are in public ownership.

St. Vincent Island (fig. 5.22) is an excellent example of dune-and-swale topography and the multiple dune ridges of a natural barrier island. The curved eastern section of the island and the southern shore have wide, sandy beaches with a history of landward retreat. At a distance of about 3.5 miles west of West Pass, a tidal inlet connecting slough and oyster pond was opened by a storm in 1889. It has since closed again. Although the beaches are wide, 80 to 100 feet, there are very low dunes along this section that make it prone to flooding during storms. To Indian Pass, most of the island is advancing seaward. St. Vincent Island is a National Wildlife Refuge, and urban-style development is not anticipated here.

Florida's barrier island chain is interrupted between Franklin County of the eastern Panhandle and southern Pasco County north of Tampa Bay, where the barrier islands again are well developed. Eight counties (Wakulla, Jefferson, Taylor, Dixie, Levy, Citrus, Hernando, and Pasco) border this great curve at the northeast edge of the Gulf of Mexico (fig. 1.1).

A significant portion of this area is underlain by limestone and fed by rivers draining limestone terrain. The result is a lack of sand, so not only are barrier islands absent, but sandy beaches and dunes are rare. The domination by rivers and creeks and the almost horizontal character of the land's slope have resulted in the marsh-swamp-delta plain that often extends for miles inland.

The geography of this coast illustrates the point made earlier that a slight rise in sea level will result in thousands of feet of shoreline retreat. In this case, flooding by the rising sea level of the last few thousands to hundreds of years has produced a vast watery transition zone between land and sea. As a result, this coastal segment of more than 200 miles in length has remained relatively untouched by man, and rightly so because it is totally unsuitable for development.

Wakulla and Jefferson counties

Almost the entire marshland-swamp coast of these 2 counties lies in the St. Marks National Wildlife Refuge. The boundary between sea and land is not always distinct, as coastal marshes give way to swamps going inland. The Sopchoppy and St. Marks rivers, plus numerous creeks, account for the marsh-swamp-delta plain that lies at sea level and rises only a few feet over several miles going inland.

The area has been flooded to considerable depths by the numerous hurricanes noted above. Southwest winds generate waves that cause significant erosion of the marsh edge, although little detailed study of such impact on this coast has taken place. The numerous marsh islands are particularly susceptible (for example, Marshes, Porter, Piney, Palmetto, Smith, Big Pass, and Sprague islands).

The refuge designation takes this extremely hazardous area out of the development market. However, several older communities do exist in the area. Panacea Park is some distance away from the coast at relatively high elevations of 10 to 15 feet. Panacea is at the head of Dickerson Bay, protected by marsh islands and a fringing marsh along the shore. The same is true for the Rock Landing–Hungry Point development. The best sites are those near the elevation 10 feet above sea level. People in these communities should be aware of the likelihood that low spots along U.S. Highway 98 and Florida Highways 30 and 61 will flood well in advance of an approaching hurricane, and early evacuation should be the rule.

Shell Point is one of the few locations along this coast with a sand beach. However, development here faces the open water of Apalachee Bay, elevations are around 5 feet, finger canals may

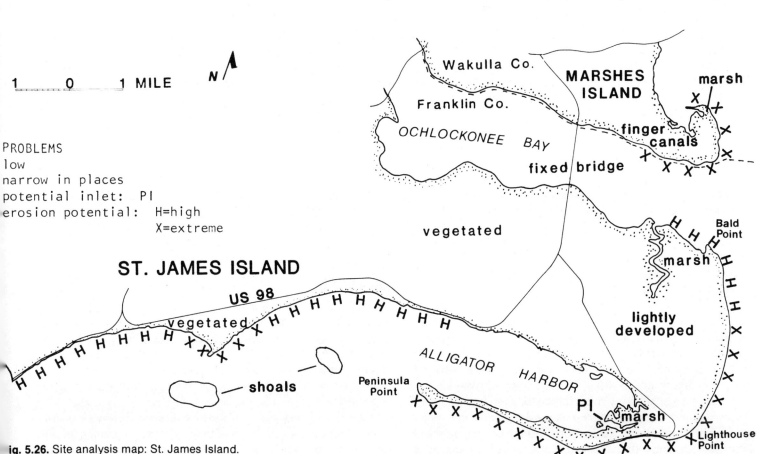

1 0   1 MILE   N

PROBLEMS
low
narrow in places
potential inlet: PI
erosion potential: H=high
                     X=extreme

Wakulla Co.
Franklin Co.
OCHLOCKONEE BAY
MARSHES ISLAND
marsh
finger canals
fixed bridge
vegetated
Bald Point
marsh
lightly developed

ST. JAMES ISLAND
US 98
vegetated
shoals
ALLIGATOR HARBOR
Peninsula Point
PI
marsh
Lighthouse Point

ig. 5.26. Site analysis map: St. James Island.

create water problems, and Florida Highway 367, the escape route, traverses a wide expanse of marsh and swamp. The same problems exist at Spring Creek at the head of Oyster Bay, although Spring Creek is somewhat protected. Live Oak Island also is in a hazardous position at the end of Florida Highway 367A.

These small developments in Wakulla County are not likely targets for extensive development since they are remote and lack the amenities that many coastal dwellers seek, namely, beaches and dunes near urban support centers. Fishermen and nature lovers may want to seek this area out, however, for its wilderness appeal. The Refuge Headquarters is located on the access road to the Coast Guard Station and Lighthouse at the mouth of the St. Marks River.

The 5 miles of Jefferson County coast is marsh, accessible only by boat. It lies entirely within the St. Marks National Wildlife Refuge. The Aucilla River forms its eastern boundary.

Taylor County

This vast marshland coast owes its origins and natural wildlife productivity to its flooded state and the fact that it is repeatedly flooded by storm events. Erosion is more severe here due to the southwest facing of the coast. This orientation is responsible for larger waves during storms. Many creeks and small rivers such as the Ecofina and Fenholloway rivers indent the shoreline. Along its southeast section there are numerous marsh islands and oyster bars. Sinkhole ponds and lakes, both within the wetlands and uplands, are evidence that this is a Karst coastline—that is, a coastline whose structure is due, in part, to the solubility of the underlying limestone. Such sinkholes are more famous for the destruction caused when they form in developed areas like the Winter Park sinkhole of a few years ago.

There are a few fishing camps off Florida Highway 361 and a few small coastal settlements such as Dekle Beach, Jug Island, and Keaton Beach at the end of Highway 361. These are older developments; nevertheless, they are in relatively high-hazard zones. They are at low elevations, facing the Gulf, with no natural protection. Future erosion is likely.

Steinhatchee lies about 1.5 miles up the Steinhatchee River on the southeast border of the county. The town's location away from the shore and on a ridge with elevations in excess of 20 feet affords some safety. Low areas along Florida Highway 51 may flood in advance of approaching hurricanes and tropical storms.

Dixie County

The extensive marsh-swamp lowland extends far inland and is pocked with sinkhole ponds. Only 3 roads traverse these wetlands to provide access to the few minor coastal developments. As in the previous areas, this region is subject to significant storm-surge flooding and erosion. Those people who choose to live in this area should be aware that it is a long drive across the flood zone in the event that evacuation is necessary.

Florida Highway 351 leads to Horseshoe Beach about a quarter of a mile inland from the beach with elevations greater than 10 feet. Some developed sites are at lower elevations. At Horseshoe

Point, houses line a shore that is eroded by waves out of the south and southwest.

A county road runs through California Swamp to the small development of Shired Island, a narrow ridge that rises to 15 feet and is surrounded by protective marsh. Cottages located on the edge of the mouth of Shired Creek are in a more precarious position.

The small communities of Suwannee and Barbree Island lie at the end of Florida Highway 349, north of the famous Suwannee River. The surrounding marsh affords protection from erosive wave attack, but the area is in the 100-year flood zone.

Levy County

The marsh coast continues south of the Suwannee River along Suwannee Sound, cut by numerous tidal creeks and extending into a wide expanse of swamp. A few isolated topographic highs, without access, exist within the swamp. Similar flooded highs have produced numerous small islands (Clark, Deer, Long Cabbage, Hog, Seabreeze, Richards) north of Cedar Key. Some of these islands have elevations of 15 feet, but they are not suitable for development.

Florida Highway 24 leads to Cedar Key, a small community just inland from the shore and with the only high ground for many miles. The elevation rises to 25 feet, and building sites above 15 feet are in the low-risk category; however, many houses are on sites below 10 feet and will be flooded even in a moderate storm. Highway 24 is adjacent to or crosses marsh and swamp, and low elevations along the highway will flood in advance of an approaching storm. Although large numbers of people would not be evacuating along this route, prudence dictates early evacuation.

Cedar Key is the gateway to Cedar Keys National Wildlife Refuge, a complex of very small islands, but some rising strikingly above sea level. Seashore Key, home of an abandoned lighthouse, rises to 52 feet. Mangrove swamps fringe some of the islets and become more common going to the south. Such mangrove fringes are important protection against the shoreline erosion that commonly attacks these islands.

East of Cedar Key the shore turns east along the north side of Waccasassa Bay, another complex of creeks, ponds, marsh, and swamp. The swamp area is known as Gulf Hammock and is cut by the Waccasassa River; it lacks development except for a few scattered hunting/fishing camps at elevations of less than 5 feet.

Yankeetown lies on the southern boundary of Levy County about 2 miles inland on the banks of the Withlacoochee River. The town boundaries extend into the bay, but the development is not on the shore. Most of the buildings are on sites with elevations of less than 10 feet and are in the flood zone. Florida Highway 40 is the lone access route to Yankeetown and Pumpkin Island, and as with other coastal routes it has low areas that will flood.

Citrus County

From the Withlacoochee River to the Crystal River the marsh-swamp zone narrows, but the shoreline continues as marsh fringe cut by many tidal creeks. South of the Crystal River is a maze of mangrove keys and marsh islands, part of which lie in the Chassahowitzka National Wildlife Refuge on Homosassa Bay and extending into Hernando County.

This coast is subject to flooding and erosion and, appropriately, is not developed. The small communities that lie along rivers or at the heads of estuaries some distance inland are often still within the 100-year flood zone.

Hernando County

South of the Chassahowitzka River is the swamp of the same name. Florida Highway 50 at the south end of the swamp provides access to small developments, the tiny Pine Island and Bayport. Pine Island is at very low elevation, while Bayport rises to 20 feet. The higher elevation plus the protective marsh fringe provides a few lower risk sites, but the area really cannot support safe development. The same is true for the limited development served by Highway 595, which approaches the coast through Weekiwachee Swamp, and near the coast in Indian Bay where a few houses along tidal creeks are in a moderate-hazard zone. Storm-surge flooding is the main threat in this area, including Bathhouse Island and Palm Island.

Pasco County

Just across the northern line of Pasco County on Florida Highway 595 is the small community of Aripeka, some distance inland from the coast and mostly above 5 feet in elevation, but still in the flood zone. Early evacuation is encouraged in the event of a hurri-cane warning. The same is true for Hudson where Highway 595 rejoins U.S. Highway 19. Many home sites are at low elevations, almost all below 10 feet, and those built along Hudson Creek to its mouth are particularly susceptible to flooding. Safer sites would be like those at Bayonet Point, approximately 1.5 miles inland at elevations greater than 10 feet.

Pasco County's shoreline is unsuitable for development as it is a continuation of the marsh coast. Locally there are fringes of reef and mangrove. The marsh zone is about a mile wide, rising over a mile or two through intermittent swamp into a sinkhole-pocked upland of about 30 feet in elevation. People seeking to locate in this area should look for inland, upland sites (away from sinkholes) at maximum elevations.

Port Richey and New Port Richey mark a cultural change in West Florida as they are the first significant developments going south through the coastal zone. The communities hint of the sub-urbanization that lies in the next county south. Although the towns are not on the shore, their position along the Pithlachascotee River places them in the flood zone. Numerous cottage sites near the river are less than 5 feet in elevation, and many are at less than 10 feet.

The shoreline of southern Pasco County also marks a signifi-cant natural change. Anclote Key lies at the northern end of an extensive barrier island chain. Anclote Key (fig. 5.27) suffers from the hazards that will be listed again and again in the coming pages, namely extreme to high rates of shoreline erosion, storm-surge flooding with wave velocity impact and overwash, the potential for new inlets to form or old inlets to migrate, and associated problems. Anclote Key National Wildlife Refuge protects people and property by preventing barrier island development on this key. The islands to the south are another story.

Pinellas County (figures 5.27–5.34)

Pinellas County has about 35 miles of beaches on barrier islands. Anclote Key (fig. 5.27), the northern tip of which is actually in Pasco County in the north, is the last of the west coast peninsular chain. North of Anclote the offshore sand layer becomes very thin with lots of rock ledges sticking through. Lack of sand and low wave energy appear to preclude the formation of barrier islands to the north until the corner is turned around the Big Bend to the northwest coast.

Indian Rocks Beach is a headland in the center of the Pinellas County chain. Wave energy tends to concentrate on headlands, and hence they are often places of severe erosion. However, erosion of the Indian Rocks headland has historically provided sand for the longshore system that has naturally nourished the islands to the north and south (fig. 5.28).

The Anclote Keys are undeveloped and have no direct connection with the mainland (fig. 5.27). These keys are low and narrow, only a few hundred feet wide, and about half of the area is intertidal mangrove. During even moderate storms with moderately associated storm tides the islands will be awash. They should remain undeveloped. Anclote Keys are separated from the next island to the south, Honeymoon Island, by a broad, open reach of shallow water beneath which lies a series of sand shoals separated by deeper channels. These features look much like barrier islands except that they are mostly underwater.

Honeymoon Island (fig. 5.27) is state parkland except for its southeastern end, which is called Dunedin Beach. As such it is in relatively good shape, again with the exception of the developed portion that has been armored with riprap in places and "protected" by construction of a jetty. Erosion has been severe in these regions.

Hurricane Pass (the name is appropriate since that is how it was formed) separates Honeymoon Island from Caledesi Island, the latter being the widest in the Pinellas barrier chain (fig. 5.29). This island is undeveloped except for several camping facilities associated with the state park that encompasses the whole island. Caledesi Island is not connected to the mainland.

Dunedin Pass separates Caledesi Island from Clearwater Beach Island to the south (fig. 5.29). Clearwater Beach is followed in turn by Clearwater Pass, Sand Key, John's Pass (fig. 5.30), Treasure Island, Blind Pass, and Long Key (fig. 5.31). With the exception of the northern end of Sand Key, which is moderately developed, these barrier islands are intensively developed with single-family dwellings, hotels, motels, and high-rise condominiums. The back sides have been extended into Clearwater and Boca Ciega bays by extensive dredge, fill, and finger canal construction. Tens of thousands of people live and vacation here.

All these islands and fill areas are low—almost everywhere under 5 feet. Few natural low dunes remain; most have been de-

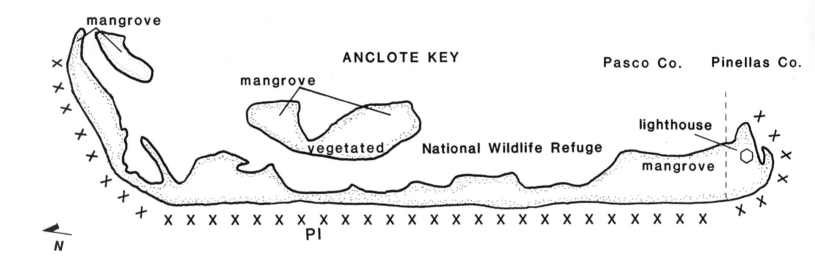

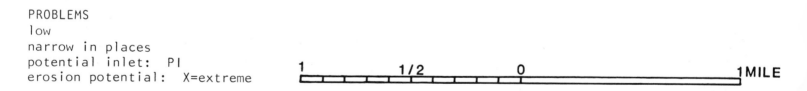

Fig. 5.27. Site analysis maps: Anclote Key and Honeymoon Island.

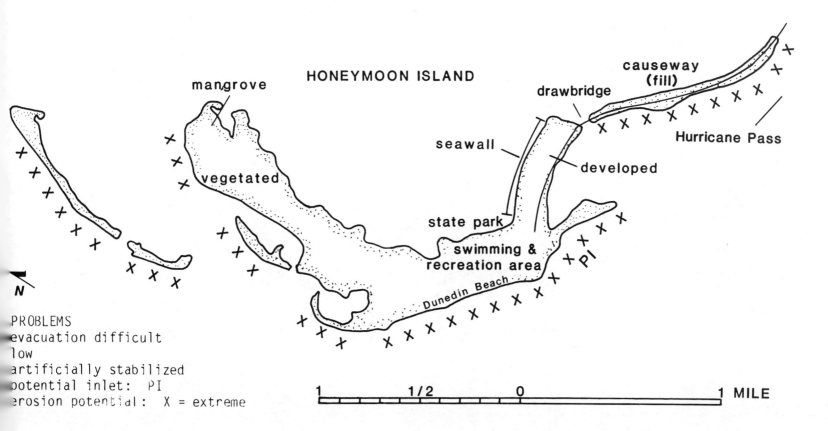

HONEYMOON ISLAND

mangrove

vegetated

causeway
(fill)

drawbridge

seawall

developed

Hurricane Pass

state park

swimming &
recreation area

Dunedin Beach

PI

N

PROBLEMS
evacuation difficult
low
artificially stabilized
potential inlet: PI
erosion potential: X = extreme

1   1/2   0   1 MILE

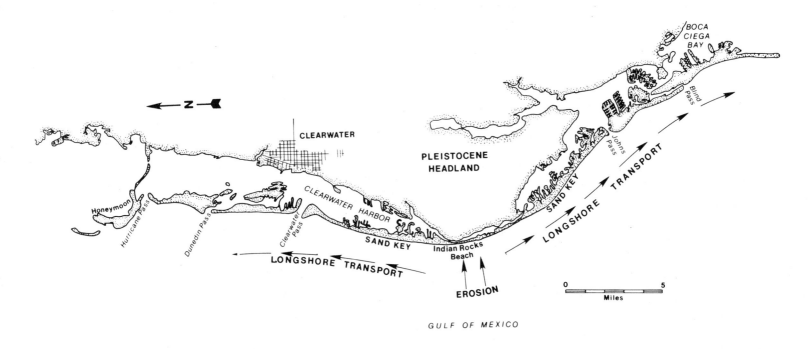

Fig. 5.28. Erosion at Indian Rocks Beach furnished sand that is carried to the islands both north and south.

stroyed. Access to the mainland is by 7 wide causeways that are low and subject to early closing in time of flooding. In addition, many of the finger canals in the bay are separated from the main section of the islands by low causeways also potentially underwater early in periods of flooding.

All these islands are subject to severe erosion and have undergone extensive engineering with many episodes of beach renourishment, jetty and groin construction, and extensive seawall development of almost every description both on the Gulf and bay sides. Blind Pass must be dredged periodically to keep it open. John's Pass also has been historically dredged.

Pass-A-Grille Channel and a series of low mangrove-dominated islands separate Long Key from Mullet Key, a triangular island at the northern edge of Tampa Bay (fig. 5.32). Fort DeSoto Park has kept development here to a minimum. Even so, the Gulf side has been extensively renourished (fig. 5.33).

Egmont Key lies squarely in the mouth of Tampa Bay (fig. 5.34). It is relatively undeveloped and is not connected to the mainland. World War II gun emplacements have been undercut by erosion, and the ruins of other facilities of that period remain. A Tampa Harbor Pilot Station is still active, located on the bay side of the island.

NOAA has calculated the 100-year storm-surge height along Pinellas County beaches to be 13.9 feet above mean sea level. A great hurricane such as Camille in 1969 or the Florida Keys hurricane of 1935 could bring a storm surge as high as 26 feet above mean sea level. Historically, hurricanes have passed within 50 miles of Pinellas beaches once every 6 to 7 years.

Storms much farther away can cause severe erosion and substantial damage. Figure 5.2 shows what a worst-case storm track might look like. Storms approaching this path actually hit the area in 1848 (details are obviously lacking) and 1921. Even a small hurricane taking this path could cause considerable damage and lead to the evacuation of hundreds of thousands of people.

Manatee County (figures 5.35–5.36)

The 12 miles of Gulf shoreline in Manatee County consist of 2 low-lying barrier islands, 7.5-mile-long Anna Maria Key (fig. 5.35) and the 4.5-mile northern section of Longboat Key (fig. 5.36). The 2 barrier islands are separated by Longboat Pass. The islands have a northwesterly-southeasterly alignment and vary in width from 400 feet near the southern end of Anna Maria to 1.25 miles near the north end of the same island. The width of Longboat Key varies from 800 feet at the south end to 3,000 feet in the middle section of the Manatee County portion. Elevations of these 2 islands are generally 5 to 9 feet, with the average being about 6 feet.

The beaches of Manatee County are composed of fine sand and shell fragments. Beach sand is probably derived from the island itself and the offshore and nearshore bottom of the Gulf. The beaches are generally narrow and steep, ranging from 40 to 60 feet in width along much of the island except at the south end of Anna Maria island where the beaches are somewhat wider. However, where seawalls and rock revetments have been placed there is no natural beach. The areas near the Passage Key and Longboat passes are characterized by rapid erosion and buildup.

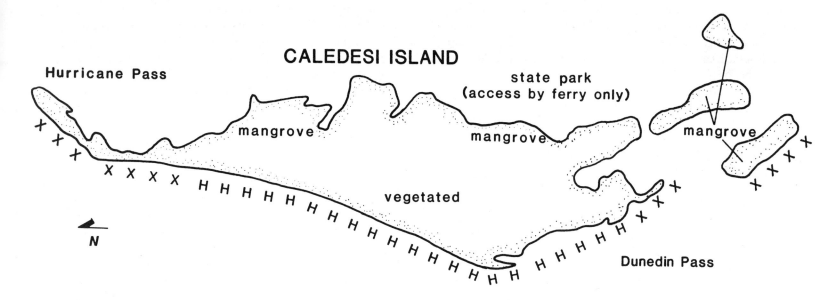

CALEDESI ISLAND

Hurricane Pass

state park
(access by ferry only)

mangrove

mangrove

mangrove

vegetated

Dunedin Pass

N

PROBLEMS
low
erosion potential: H=high
                     X=extreme

Fig. 5.29. Site analysis map: Caledesi Island and Clearwater Beach Island.

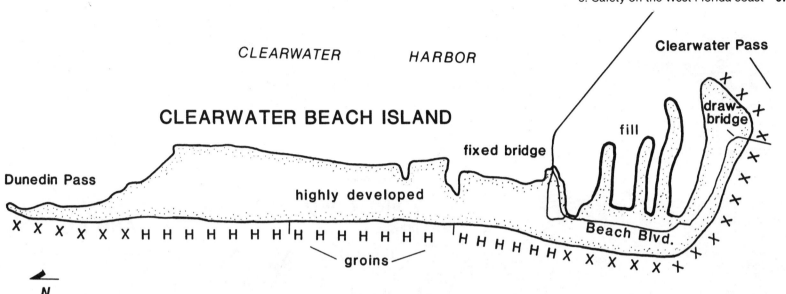

CLEARWATER   HARBOR

CLEARWATER BEACH ISLAND

Clearwater Pass

draw-bridge

fill

fixed bridge

Dunedin Pass

highly developed

Beach Blvd.

X X X X X X X H H H H H H H H H H H H H H H H H H H H H H H H H X X X X X X X X X X

groins

N

PROBLEMS
evacuation difficult
low
artificially stabilized
erosion potential: H=high
                    X=extreme

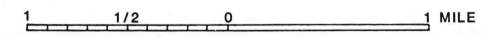

1   1/2   0   1 MILE

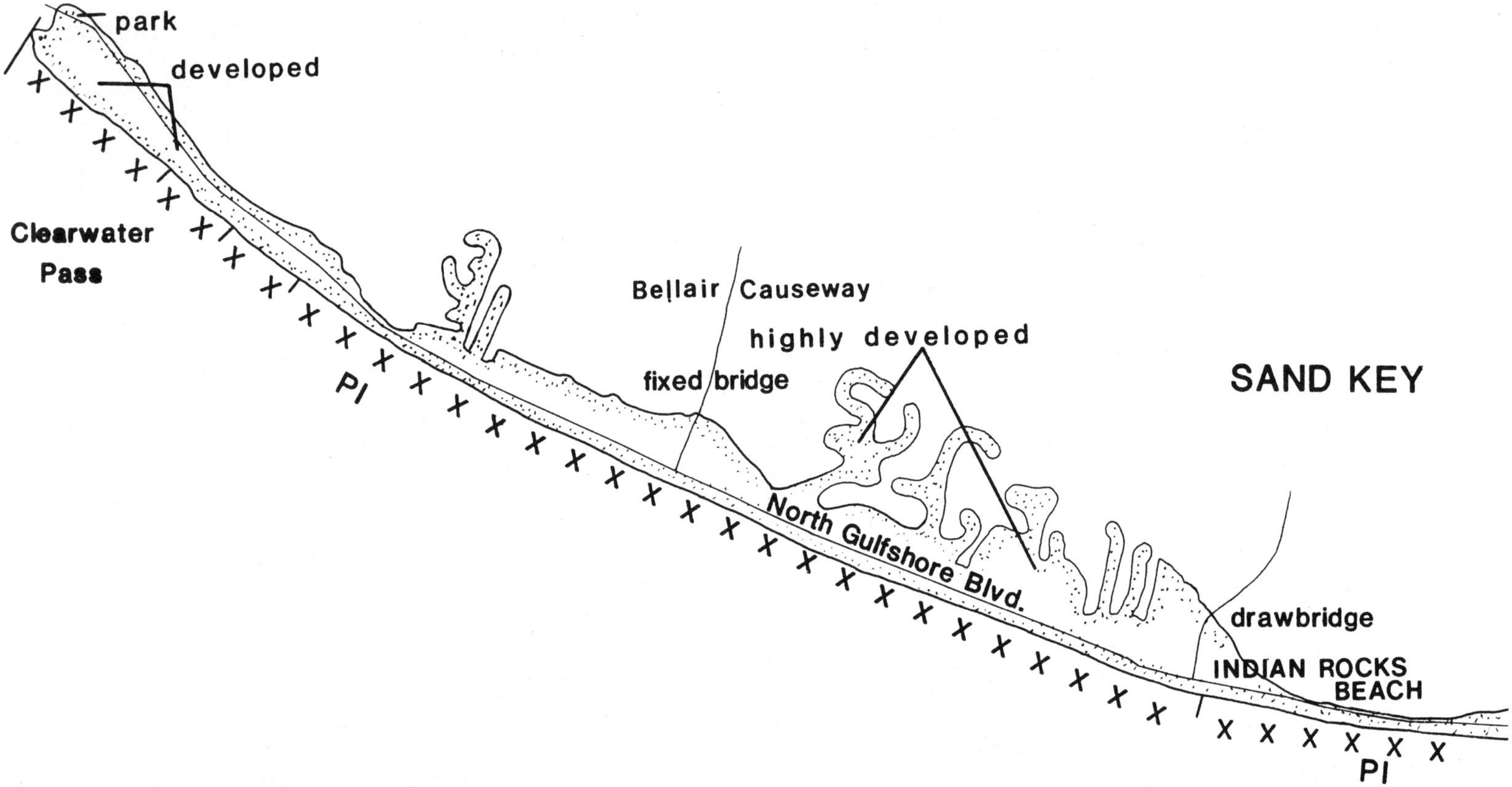

park

developed

Clearwater
Pass

Bellair Causeway

highly developed

fixed bridge

SAND KEY

Pl

North Gulfshore Blvd.

drawbridge

INDIAN ROCKS
BEACH

Pl

Fig. 5.30. Site analysis map: Sand Key, Clearwater Pass, through John's Pass.

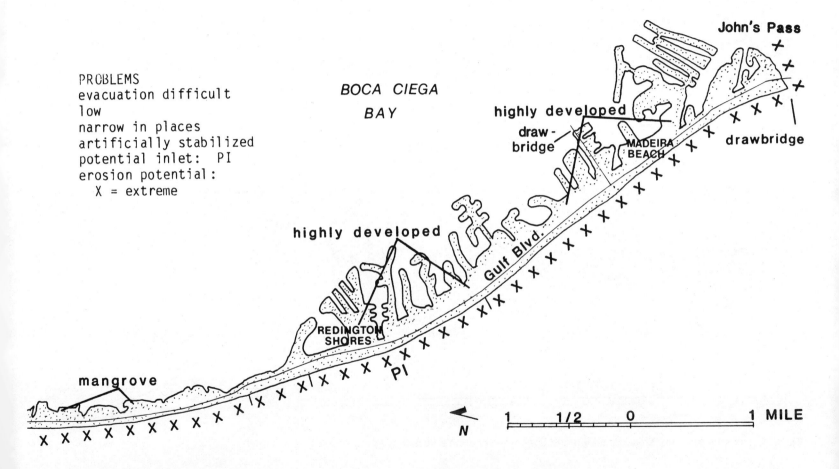

PROBLEMS
evacuation difficult
low
narrow in places
artificially stabilized
potential inlet: PI
erosion potential:
  X = extreme

BOCA CIEGA
BAY

John's Pass

highly developed

draw-
bridge

MADEIRA
BEACH

drawbridge

highly developed

Gulf Blvd.

REDINGTON
SHORES

PI

mangrove

1   1/2   0   1 MILE

N

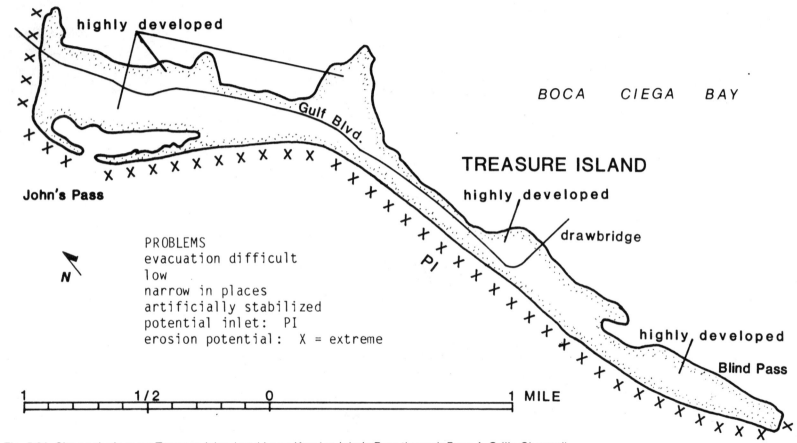

highly developed

BOCA CIEGA BAY

Gulf Blvd.

TREASURE ISLAND

highly developed

drawbridge

John's Pass

PI

PROBLEMS
evacuation difficult
low
narrow in places
artificially stabilized
potential inlet: PI
erosion potential: X = extreme

highly developed

Blind Pass

N

1   1/2   0   1 MILE

Fig. 5.31. Site analysis map: Treasure Island and Long Key (or John's Pass through Pass-A-Grille Channel).

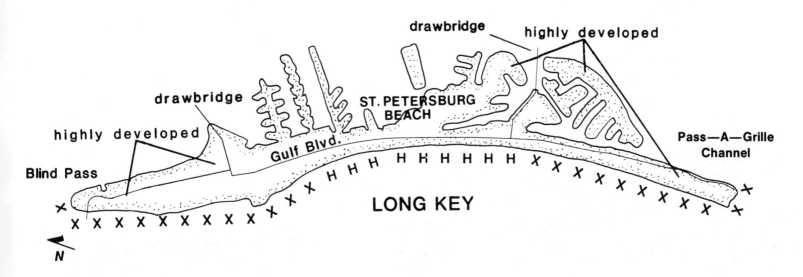

drawbridge   highly developed

drawbridge

highly developed

ST. PETERSBURG
BEACH

Blind Pass

Gulf Blvd.

Pass—A—Grille
Channel

X

X X X X X X X X X X X X X H H H   H H H H H H H   X X X X X X X X

LONG KEY

X X X X X X X X X X

X X X X X X X

N

PROBLEMS
evacuation difficult low
artificially stabilized erosion
potential: H = high
            X = extreme

1   1/2   0   1 MILE

Living with the West Florida shore

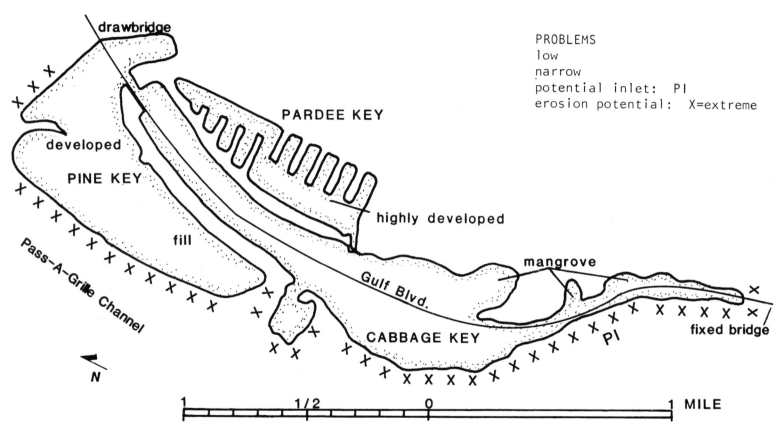

Fig. 5.32. Site analysis map: Pardee Key and Cabbage Key.

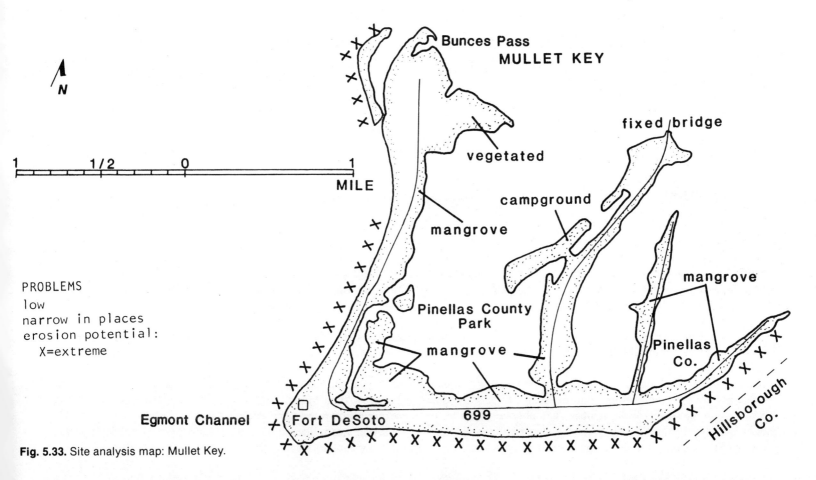

Fig. 5.33. Site analysis map: Mullet Key.

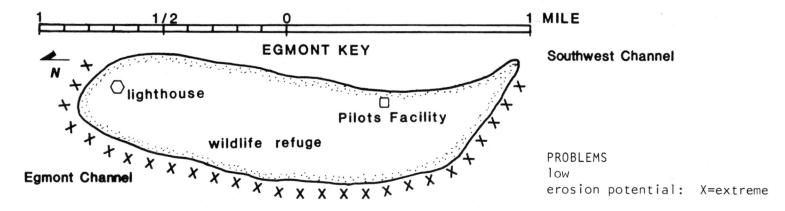

Fig. 5.34. Site analysis map: Egmont Key.

Since 1900 a total of 30 known hurricanes and tropical disturbances have passed within a 50-mile radius of the Manatee County shoreline. Of that total, 14 were classified as hurricanes and 16 as lesser storms. The relative frequency of hurricanes and tropical disturbances affecting Manatee County is about 1 in 2.5 years. Some of the more remarkable hurricanes and storms that affected Manatee County include the hurricane of October 1910 and the hurricane of October 1921 that caused tidal flooding 7 feet above normal and covered the northern end of Anna Maria with 5 feet of water. The hurricane of September 1926 flooded much of the barrier islands and caused $1.0 million in damages from flooding in the Bradenton area. A severe tropical storm in 1935 undermined and damaged the entire reach of Manatee County beaches. The October 1946 hurricane passed almost directly across Manatee County beaches, flooding all low-lying beaches and causing significant property damage. The September 1950 hurricane caused almost complete flooding of Anna Maria Key, 15 to 20 feet of beach retreat, and cut through the beach road at several locations. Hurricane Gladys in October 1968 caused 40 feet of beach erosion on the north end and 20 feet of erosion along the central Gulf beach of Anna Maria. The "No Name Storm" in June 1982 caused severe beach retreat and damaged recently nourished beach. The 100-year storm tides in Manatee County are estimated at 13.5 feet above mean sea level, excluding wave height. As of August 1981 there were 10,549 flood insurance policies valued at more than $577 million in effect in Manatee County. Much of this coverage is located along the coastal areas of the beach and the bay. In the event a major hurricane strikes this area, property damage could run into hundreds of millions of dollars.

The interaction of winds, waves, tides, and currents in addition to sea-level rise and periodic storms produce the changes in the beaches. The prevailing winds are from the northeast and north during the winter and east and south during the remainder of the year. The tides in the Manatee County area are a mixture of twice-daily and daily types with a mean daily range of 2.3 feet. NOAA estimates that 10- and 25-year storm-surge levels at Anna Maria will be 5.6 feet and 8.9 feet above mean sea level, flooding most of the barrier islands. Currents are predominantly tidal with average maximum and minimum at Longboat Pass 3.0 and 1.7 feet per second. The movement of sand on the beach diverges near the center of Anna Maria Key—net littoral drift is northerly in the area north of Holmes Beach and southerly south of Holmes Beach. Approximately 1.5-mile sections near the inlet are influenced by the inlets.

The shoreline has a history of advance and retreat. According to the U.S. Army Corps of Engineers, the northern two-thirds of the Gulf shoreline of Anna Maria Key advanced an average of 210 feet from 1875 to 1968, while the southerly one-third retreated 422 feet on average from 1833 to 1968. The northerly half of Longboat Key in Manatee County advanced an average of 355 feet from 1883 to 1968, while the southern half retreated an average of 173 feet during this period. From 1946 to 1968 the entire shoreline retreat has continued at most locations along the Manatee County beaches. The offshore 6-, 12-, and 18-foot depth contours generally have moved landward between 1926 and the present. This means a net erosion trend for the system.

Individuals and governments in Manatee County have installed various structural solutions to protect their property and facilities and restore the beaches. These include seawalls, revetments, groins, jetties, and artificial nourishment. These structures have temporarily slowed shoreline retreat but have proved ineffective in reducing long-term beach erosion. Longboat Pass is stabilized and maintained at a 12-foot depth and 150-foot width from Gulf Pass to Longboat Pass Bridge, and at a 10-foot depth and 100-foot width from Longboat Pass Bridge to Cortez Bridge. The total authorized project length (September 1980) was 2.6 miles at a cost of $1.24 million. The renourished beaches require frequent maintenance as became evident after the "No Name Storm" of June 1982. The most recent project is the nourishment of 3.9 miles of beach at 3 sections of Anna Maria Beach. The proposal calls for 1.3 million cubic yards of sand for a 100-foot-wide beach at an estimated cost of $7.1 million.

Sarasota County (figures 5.36–5.40)

Sarasota County has 35 miles of barrier island beach and encompasses the southern half of Longboat Key. Longboat Key is moderately developed on the Gulf side and has considerable finger canal development on the southern end facing Sarasota Bay (fig.

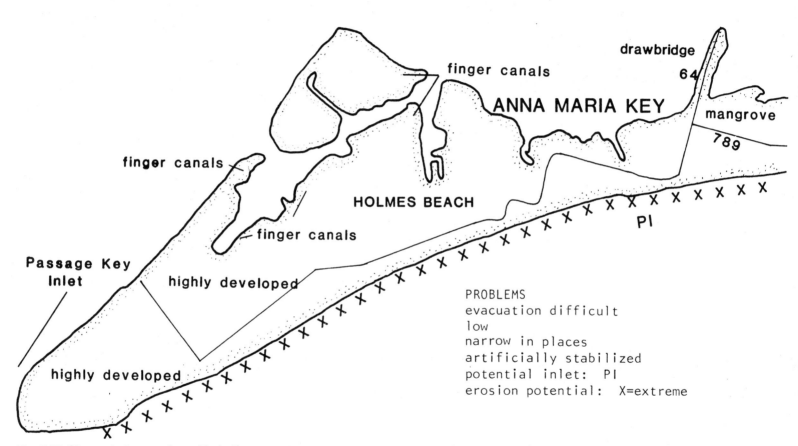

finger canals

drawbridge

64

ANNA MARIA KEY

mangrove

789

finger canals

HOLMES BEACH

finger canals

PI

Passage Key
Inlet

highly developed

PROBLEMS
evacuation difficult
low
narrow in places
artificially stabilized
potential inlet: PI
erosion potential: X=extreme

highly developed

Fig. 5.35. Site analysis map: Anna Maria Key.

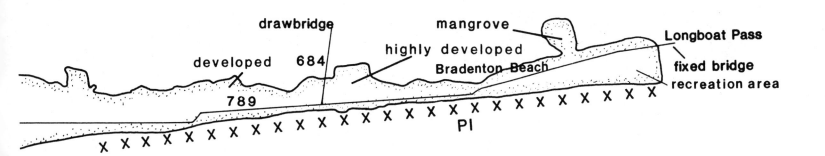

developed

drawbridge

684

789

mangrove

highly developed
Bradenton Beach

Longboat Pass

fixed bridge

recreation area

PI

N

1   1/2   0   1 MILE

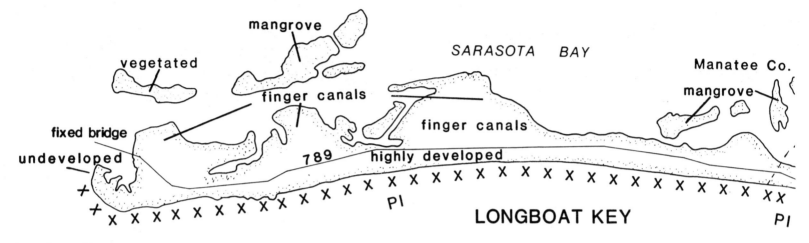

Longboat Pass

```
PROBLEMS
evacuation difficult        potential inlet:  PI
low                         erosion potential:
narrow in places              X = extreme
artificially stabilized
```

Fig. 5.36. Site analysis map: Longboat Key.

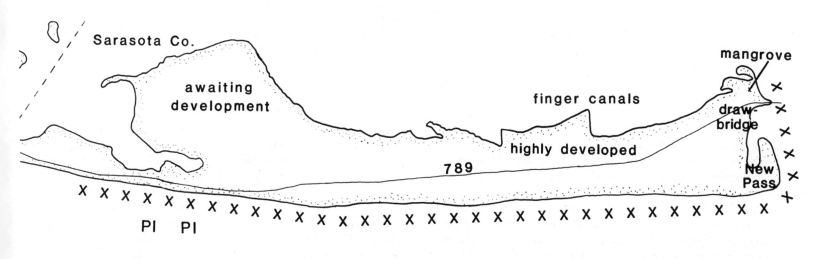

Sarasota Co.

awaiting
development

mangrove

finger canals

draw-
bridge

highly developed

789

New
Pass

X X X X X X X X X X X X X X X X X X X X X X X X X X X X

Pl   Pl

N

1   1/2   0   1 MILE

5.36). New Pass separates Longboat Key from Lido Key (fig. 5.37), with most development here on the bay side and on St. Armands Key at the end of the long causeway to Sarasota.

Big Sarasota Pass lies between Lido Key and Siesta Key (fig. 5.38). Siesta Key developed with the usual canals, fill, and beach engineering projects. Midnight Pass separates Casey Key to the south from Siesta Key. Casey Key is a low, narrow, developed barrier island (fig. 5.39). In most places it is only a few hundred feet wide and less than 15 feet high.

Sarasota County can expect a storm surge from the 100-year storm of at least 12.6 feet above mean sea level.

The beaches of Sarasota County are composed of fine sand and shell fragments. They are generally narrow and steep along much of the shoreline—with the exception of the public beaches at Lido Key, Siesta Key, and Casey Key where beach nourishment or construction of jetties helped to provide wider beaches. The kinetic energy associated with winds, waves, currents, and tides shape and reshape the beaches. The predominant winds are from the northeast and north during the winter and east and southeast during the rest of the year. The tides in Sarasota County are a mixture of twice-daily and daily. The range of the tide is 2.3 feet along the Gulf and 2.1 feet in Sarasota Bay.

Since 1900 a total of 32 hurricanes and tropical storms have passed within 50 miles of the Sarasota County shoreline. The probability of hurricanes striking Sarasota County beaches is 1 in 6.5 years, and hurricanes and tropical storms combined 1 in 3 years. The most recent storm to affect Sarasota County was the "No Name Storm" of June 1982, which caused severe beach erosion and flooding on various islands. As of August 1981 there were 16,467 subsidized flood insurance policies valued at $890 million in effect in Sarasota County. During a major hurricane, property damage could run into billions of dollars.

Shoreline changes indicate the erosion and buildup trends along the county coastline. From 1883 to 1967 Longboat Key retreated an average of 242 feet. Lido Key shoreline advanced at the north and south ends and retreated in the central section from 1939 to 1967. These changes were influenced by the placement of fill from New Pass dredging. The northern end of Siesta Key built up an average of 602 feet from 1883 to 1967, while the southern half retreated an average of 153 feet during the period from 1883 to 1953. The middle 1-mile section of Siesta Key retreated an average of 108 feet from 1883 to 1967. The northern Casey Key shoreline retreated 225 feet from 1883 to 1939, while the remainder advanced slightly. From 1939 to 1953 Casey Key's entire shoreline retreated an average of 49 feet. This trend continues today. In almost all cases the shoreline stretches at the mouths of inlets have fluctuated back and forth much more than those away from them. In recent years shoreline south of Big Sarasota Pass and north of Midnight Pass have retreated severely, but precise data are lacking.

The offshore 6-, 12-, and 18 foot depth contours generally retreated off Longboat Key, while they advanced and retreated in about equal amounts along Siesta Key. The 6-foot depth contour off Lido Key receded an average of 690 feet along the north end and advanced 596 feet in the southern half during 1883 to 1967.

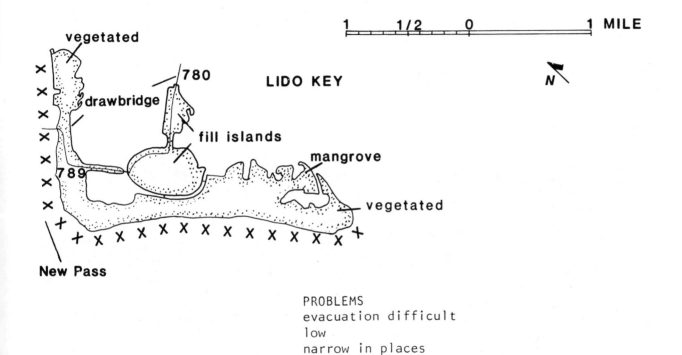

vegetated

780

LIDO KEY

N

drawbridge

fill islands

mangrove

789

vegetated

New Pass

PROBLEMS
evacuation difficult
low
narrow in places
erosion potential: X=extreme

Fig. 5.37. Site analysis map: Lido Key.

Scale: 1 1/2 0 1 MILE

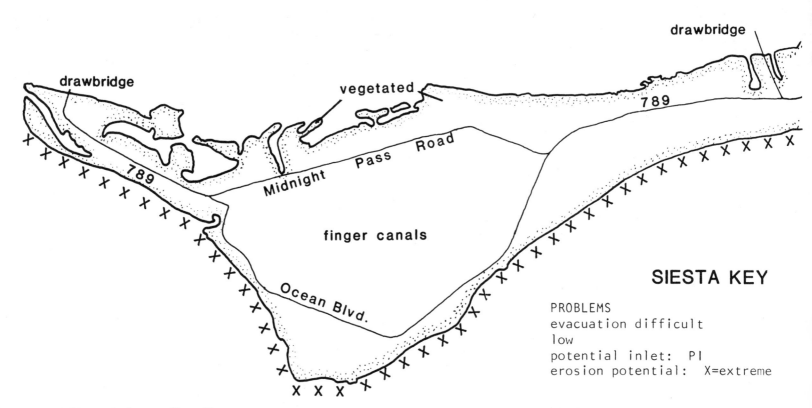

drawbridge

drawbridge

vegetated

789

789

Midnight Pass Road

finger canals

Ocean Blvd.

SIESTA KEY

PROBLEMS
evacuation difficult
low
potential inlet: PI
erosion potential: X=extreme

Fig. 5.38. Site analysis map: Siesta Key.

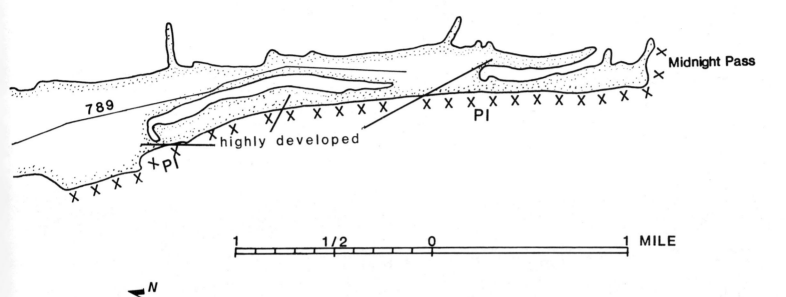

789

X X X X X X highly developed X X P X

X X X X

P X X X X X X X X X X X X X X X X X X X X X X Pl

Midnight Pass

1 1/2 0 1 MILE

N

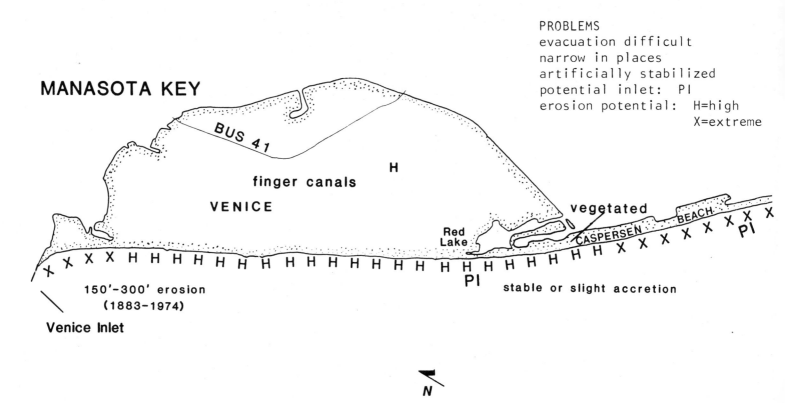

PROBLEMS
evacuation difficult
narrow in places
artificially stabilized
potential inlet: PI
erosion potential: H=high
X=extreme

MANASOTA KEY

BUS 41

finger canals

VENICE

H

vegetated

Red Lake

CASPERSEN BEACH

PI

150'-300' erosion
(1883-1974)

PI

stable or slight accretion

Venice Inlet

N

Fig. 5.39. Site analysis map: Casey Key and northern Manasota Key.

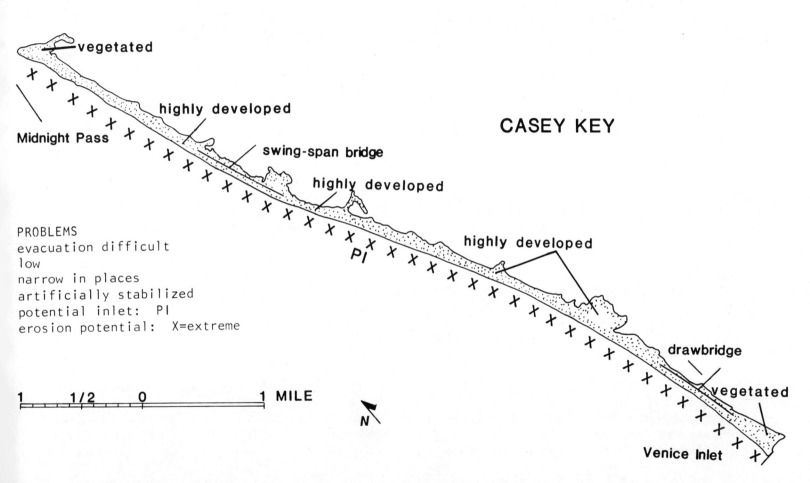

vegetated

Midnight Pass

highly developed

swing-span bridge

highly developed

CASEY KEY

highly developed

PI

PROBLEMS
evacuation difficult
low
narrow in places
artificially stabilized
potential inlet: PI
erosion potential: X=extreme

drawbridge

vegetated

1   1/2   0   1 MILE

N

Venice Inlet

During this period 12- and 18-foot contours advanced 460 feet and 434 feet, respectively, along the northerly portion and retreated a net average of 208 and 408 feet, respectively, along the southerly portion of Lido Key. At Casey Key the 6- and 12-foot depth contours off the northern and southern ends of the island retreated, while they advanced along the middle of the island between 1883 to 1953. During the same period the 18-foot depth contour advanced along the northerly half and receded along the southerly half of the island. At the present time the shorelines along Casey Key have steep offshore profiles and almost no beach.

Various methods to protect property and control erosion along the Sarasota County shoreline have included seawalls, bulkheads, revetments, groins, jetties, sand bags, and beach nourishment. Lido Key has been nourished from an inlet maintenance dredging operation at a cost of $3.7 million. There are proposals for nourishment of other beaches.

New Pass was created by a hurricane in September 1848 that breached Longboat Key. The pass is now maintained at a depth of 10 feet and a width of 100 feet by the Corps of Engineers. Venice Inlet also is maintained by the Corps of Engineers as part of the Intracoastal Waterway system.

Venice Inlet to south Sarasota County line (Manasota Key barrier spit)

The beaches south of Venice Inlet are narrow and backed by low berms and some dunes south of Caspersen Beach (figs. 5.39 and 5.40). Manasota Key has some of the highest elevations on the southwest Florida Coast—averaging 10 to 12 feet and in a few places exceeding 16 feet. However, this peninsula also is one of the narrowest, ranging in width from less than 100 feet to no more than 1,500 feet. From the south end of Tarpon Center Drive to Red Lake the beaches are part of the mainland.

Beaches are composed of fine quartz sand and shell fragments. South of the Venice jetty the beach is 40 to 60 feet wide with steep scarps cut by wave action. There are no beaches remaining where seawalls and groins or revetments have been placed. The width of the beach increases to about 100 feet at the pier by the public beach. In general, the width varies from 30 to 75 feet for the entire southern section.

Manasota Key has a history of erosion and some buildup in the middle section. An inlet (Casey Pass) existed 1,000 to 1,500 feet north of the present jettied inlet. It was plugged when the present Venice jetty was built. This jetty acts as a barrier for sand carried by longshore currents. The beach at Casey Key north of Venice Inlet has advanced, while that south of the inlet has retreated. From 1883 to 1974 the shoreline from Venice Inlet to Red Lake eroded 150 to 350 feet; from Caspersen Beach to Manasota Beach the shoreline remained stable or built out slightly in the north section while eroding slightly near Manasota Beach. The presence of steep scarps and offshore slopes are an indication of present-day general erosion.

Since 1900 more than 32 hurricanes and tropical storms have passed within 50 miles of the coast, with a frequency of 1 storm every 4 years and 1 hurricane every 9 years. The 100-year storm tides are estimated at between 11.5 feet and 12.5 feet above mean sea level—excluding wave heights. Implications for flood damage are obvious for the barrier spit since it is generally below 10 feet in elevation.

Emplacement of seawalls, revetments, and groins to stop natural island migration has not stopped erosion but in many instances accentuated the problem by steepening offshore depth profiles. This problem is particularly critical between the Venice jetty and Venice public beach to the south. The placement of dredged materials from the inlet and Intracoastal Waterway maintenance program has provided some temporary relief to the beaches south of the Venice Inlet—but shoreline retreat continues.

Charlotte County (figures 5.40–5.41)

Charlotte County has about 14 miles of barrier islands and spits. They are the southern 4 miles of Manasota Key, also known as Punta Gorda Beach and Englewood Beach (fig. 5.40); Knight–Don Petro–Little Gasparilla Island, 8 miles of barrier islands that were separated by Bocilla Pass, Blind Pass, and Little Gasparilla Pass in recent times (fig. 5.40); and the northern 2 miles of Gasparilla Island (fig. 5.41). These barrier islands range in width from 80 feet to 2,000 feet and have elevations ranging from 5 to 9 feet, with the average being slightly above 5 feet.

The beaches, composed of quartz sand and a large portion of broken or whole shells, are generally narrow and steep; offshore profiles also are steep. At Manasota Gardens north of Englewood Beach the beach is 30 feet wide. Opposite the Manasota Bridge and south to the Manasota Beach Club the beach is about 70 feet wide. At Englewood public beach the beach is 50 to 60 feet wide in fair weather, but there is no beach during rough weather. The back shore in this area and south to Stump Pass is less than 5 feet in elevation. Charlotte Beach State Recreation Area has a few very low and narrow sections that are overwashed during winter storms.

South of Stump Pass to Gasparilla Pass the beaches are very dynamic due to the presence and influence of 2 open and 4 closed inlets. The beach width along the entire section ranges from 50 to 70 feet with backshore elevations of generally less than 5 feet. Don Pedro Island has extensive filled areas in subdivisions.

Based upon historic data, maximum tides are estimated to be about 11.5 feet above mean sea level for stillwater in Charlotte County. During storms, wave heights would be added to this level.

Other than catastrophic hurricanes, winter weather produces most of the adverse conditions causing serious erosion and overwash. Since 1900 a total of 30 known hurricanes and tropical disturbances have passed within a 50-mile radius of the county's shoreline. Of this total 14 storms were classified as hurricanes. The frequency of hurricanes for the period 1900–1976 is 1 in 5 years. The frequency of hurricanes and tropical storms is 1 in 2.5 years for the same period—a relatively high rate. Memorable hurricanes that affected the Charlotte County shoreline and coastal property include the October 1921 hurricane that produced 7- to 11-foot flood levels along the coast and 8-foot flood levels at Punta Gorda; the September 1926 hurricane, one of the most destructive in Florida's history, produced 11- to 12-foot tides and opened Redfish Pass on Captiva Island. The hurricane of September 1935

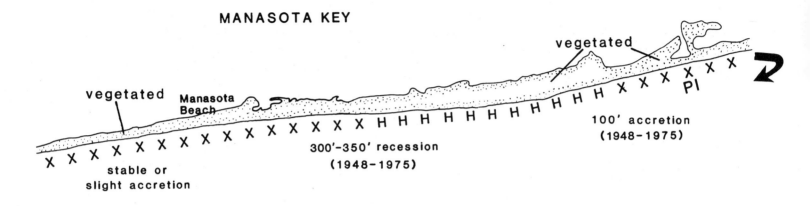

MANASOTA KEY

vegetated

vegetated
Manasota
Beach

100' accretion
(1948–1975)

300'–350' recession
(1948–1975)

stable or
slight accretion

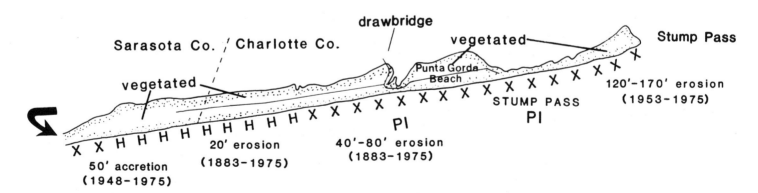

Sarasota Co. / Charlotte Co.

drawbridge

vegetated

Stump Pass

vegetated

Punta Gorda
Beach

120'–170' erosion
(1953–1975)

STUMP PASS
PI

Pl

20' erosion
(1883–1975)

40'–80' erosion
(1883–1975)

50' accretion
(1948–1975)

Fig. 5.40. Site analysis map: Manasota Key through Little Gasparilla Island.

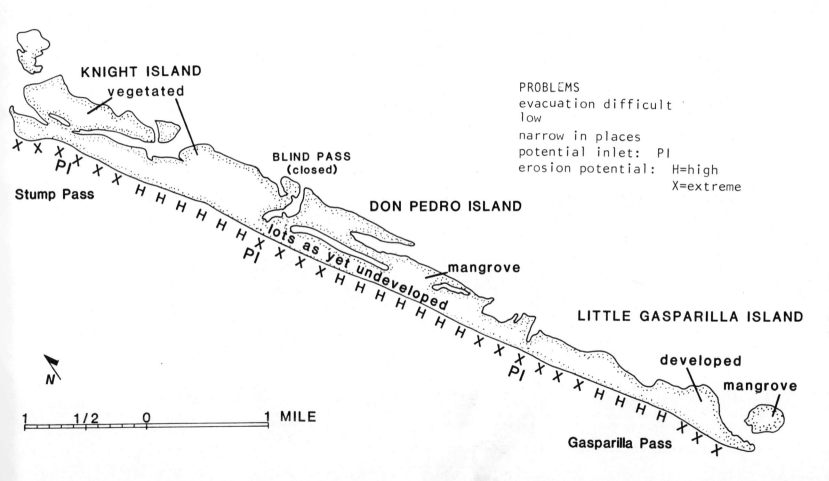

KNIGHT ISLAND
vegetated

PROBLEMS
evacuation difficult
low
narrow in places
potential inlet: PI
erosion potential: H=high
                    X=extreme

BLIND PASS
(closed)

DON PEDRO ISLAND

Stump Pass

lots as yet undeveloped

mangrove

LITTLE GASPARILLA ISLAND

developed

mangrove

N

Gasparilla Pass

1   1/2   0   1 MILE

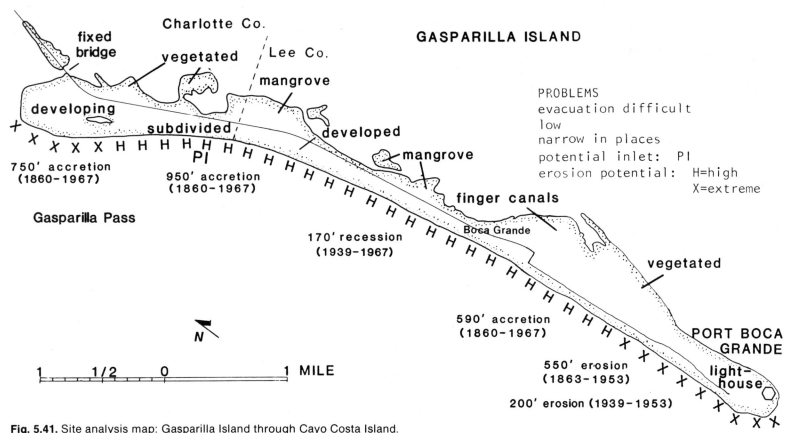

Fig. 5.41. Site analysis map: Gasparilla Island through Cayo Costa Island.

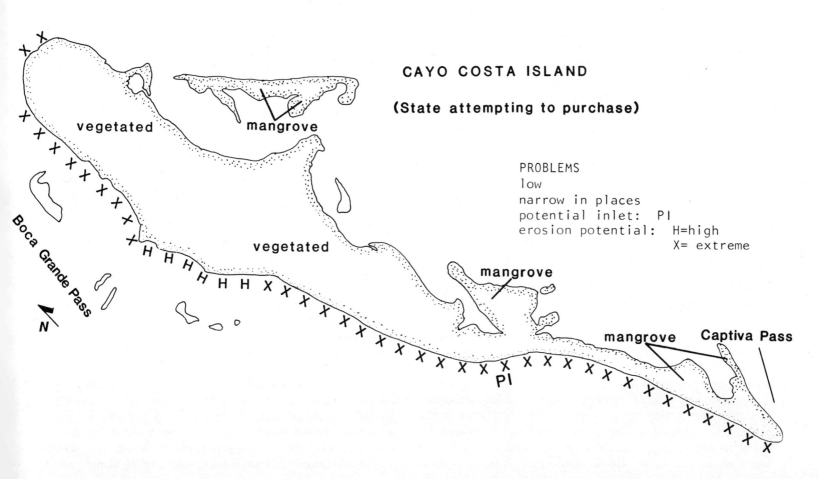

CAYO COSTA ISLAND

(State attempting to purchase)

PROBLEMS
low
narrow in places
potential inlet: PI
erosion potential: H=high
                     X= extreme

vegetated

mangrove

vegetated

mangrove

mangrove

Captiva Pass

Boca Grande Pass

N

H H H H H H

PI

was another large storm with wave heights estimated at 16 feet. The barrier islands were flooded to a depth of several feet. The hurricane of October 1944 had 100-mph winds, and its 7-foot tides overtopped Gasparilla Island, eroding 50 to 60 feet of beaches in places. Hurricane Donna (September 1960), one of the great storms of the century, caused severe erosion damage and flooding. The total damage in Florida caused by Donna exceeded $87 million (1960 prices), and over $26 million of those damages were between the Collier County and Sarasota County coast. Tides of 4 to 6 feet above normal topped the barrier islands in several places, cutting through the narrow beaches to the bay. Hurricane Alma of June 1966 created tides of 5 feet above normal at Fort Myers and caused much erosion and flooding along the islands. Hurricanes Frederic and David in 1979 caused higher than usual tides, overwash, and erosion in the state park area and in the Knight–Don Pedro–Little Gasparilla Island chain.

According to the U.S. Army Corps of Engineers' 1979 report, future damage to the barrier beaches "could be significantly greater than in the past." This is mainly due to instability and recession of the beaches as a result of slowly rising sea level—about 1 foot in the last 100 years—periodic storms, and increased development. Also the beaches have high shell contents, are generally coarse-grained, and have steep profiles.

The northern 2 miles of Charlotte County beach have withdrawn up to 100 feet during the last century. The Englewood public beach area, on the other hand, has remained stable or even built out 20 to 40 feet in the last 100 years. The southern mile of the beach has been extremely volatile and dynamic. It eroded by 50 to 170 feet during the period from 1953 to 1975. Stump Pass has migrated more than 1.3 miles south since 1883. The entire Charlotte State Recreation Area property is subject to periodic overwash, flooding, and erosion. A 1944 storm opened a 400-foot wide inlet, 3,500 feet north of present Stump Pass.

Knight–Don Pedro–Little Gasparilla Island has been cut by at least 5 different inlets at different times during the 1883–1981 period. The inlets that have closed from north to south are Bocilla Pass on Knight Island, an unnamed inlet on Knight Island, Blind Pass between Knight and Don Pedro Island, and Little Gasparilla Pass between Don Pedro and Little Gasparilla islands. Generally, the beach areas one-half to 1 mile north and south of inlets are the most dynamic of all on barrier islands and must be considered high-hazard zones for any structures. Low elevations make the island vulnerable to flooding.

The northern 2 miles of Gasparilla Island in Charlotte County are characterized by both localized buildup and erosion. The area at the mouth of Gasparilla Pass has generally built out and has parallel dune ridges. For about 1 mile the beach is about 50 feet wide and the toe of the beach ridge is 4.5 feet in elevation. Due to narrow width and low elevation the rest of the beach has a high flood hazard.

Of the county's 14 miles of Gulf shoreline, only 4 miles of the northern section on Manasota barrier spit and 2 miles of the southern section are accessible by automobile to the public. There is only one county beach, Englewood Beach (1,650 feet and 11 acres), to serve the existing population of 56,000. The state of Florida purchased the southernmost 6,000 feet of Manasota Key with a total of 212 acres for $888,000 in May 1971 to develop as a state beach recreation area. No development or improvements have taken place as of this writing. Charlotte County's population is expected to increase to 123,000 by the year 2000 and 227,000 by the year 2030. The county has no existing plan to acquire beach front to augment the recreation needs of the present and future population.

On the northern 4 miles of the county's beaches property owners have placed 480 feet of rock groins, 400 feet of Budd groins, 980 feet of concrete seawall, and 3,340 feet of rock revetment. According to the U.S. Army Corps of Engineers, those "local efforts have not been successful in retarding landward movement of the mean high water line." However, it should be noted that the Corps as well as the state Department of Natural Resources continue to grant permits for similar structures to applicants. There have been unsuccessful localized beach fill projects as well. Interestingly, the Corps of Engineers itself has proposed a beach "restoration and nourishment" project for the northern 3.3 miles of beach. The project recommends placement of 300,000 cubic yards of sand along 3.3 miles from the Sarasota County line to about 3,000 feet north of Stump Pass. Annual "renourishment" of 55,000 cubic yards will be required under the proposal. It is designed to provide a 30-foot wide beach at an elevation of 5 feet relative to mean low water or 3 or 4 feet above mean sea level. For the southern 6,000 feet of the state recreation area, the plan proposes an artificial dune of 9 feet, 20 feet wide at the top, with 5 dune crossovers at intervals of 1,000 feet. A 550-foot terminal groin at the south end

of the park and north of Stump Pass, with a top elevation at 5.5 feet at the landward end, also is proposed along with a 700-foot rubble mound revetment tied to the terminal groin and extending north, with a top elevation of 5.0 feet. The Corps of Engineers estimates a project life of 50 years. The cost for the 3.3-mile section is estimated at $3.9 million initially and $208,000 in annual maintenance and renourishment costs. The costs for the state park area are estimated at $1.043 million initially and $17,000 in annual maintenance and renourishment costs. It should be noted that no beach nourishment or erosion control project lasts for 50 years. It is recommended that sand supply for the project be derived from a borrow site about 4,000 to 6,000 feet west of Stump Pass in the Gulf. It is interesting to note that the costs of protecting the state park exceed considerably the purchase costs for the property.

During 1980–81 the U.S. Army Corps of Engineers dredged the Stump Pass channel 8 feet deep and 150 feet wide for navigation uses at an initial cost of $522,000 and an annual operating and maintenance cost of $40,000. Approximately 97,000 cubic yards of spoil was placed on the state recreation area beach covering about 10 acres of grassbeds.

This example illustrates that public purchase of the remaining undeveloped barrier is much more economical than subsidized urban development. The barrier islands and beaches, once acquired for public use, should be used as natural recreation areas, without efforts to "stabilize" them as proposed for the Charlotte State Recreation Area. "Stabilization" is costly as well as ineffective.

Lee County (figures 5.41–5.46)

The 44-mile coastline of Lee County is made up of a group of low barrier islands that generally follow a south-southeasterly trend. Sanibel Island is an exception, trending in an east-north-easterly direction.

From north to south Lee County has 9 inlets—Boca Grande Pass, Captiva Pass, Redfish Pass, Blind Pass (presently closed), San Carlos Bay and Matanzas Pass, Big Carlos Pass, New Pass, Little Carlos Pass, and Big Hickory Pass—bordering the barrier islands. Only Boca Grande Pass, San Carlos Bay and Matanzas Pass, Big Hickory Pass, and perhaps New Pass are dredged and maintained by the Army Corps of Engineers. Captiva, Redfish, Blind, and Little Hickory passes are natural passes at this time.

Big Hickory Island (behind Bonita Beach) is separated from the mainland by a narrow bay about 2,000 feet across, but most of the islands are separated from the mainland by bays which are between 1.5 to 7 miles wide.

Most of the islands are less than 8 feet high and lack pronounced dunes. An exception is Sanibel, which has a field of multiple, bow-curved dunes, many of which have been leveled for urban development in the past 15 to 20 years.

Low elevations combined with lack of dunes and extensive engineered beaches make Lee County's barrier island chain especially susceptible to storm damage. NOAA estimates the 100-year stillwater flood conditions on Lee County's barriers at 13.6 feet, more than 5.5 feet above the 8-foot elevation that is maximum on most of the islands. Remember that storm surge could increase this mark, and storm waves also would be added.

Since 1900 a total of 35 hurricanes and tropical storms have passed with a 50-mile radius of the Lee County coastline. Of these, 19 were hurricanes and 16 tropical storms. Available records suggest the frequency of hurricanes is 1 in 4 to 6 years, and the frequency for all tropical storms is 1 in 2 to 3 years.

The following descriptions of the most damaging and memorable hurricanes to strike Lee County are informative. In October 1873, a storm destroyed Punta Rassa and produced a tide 14 feet above mean sea level. In October 1910 a hurricane produced 90-mph winds, tides 10.5 feet above mean sea level at Everglades and 14.3 feet at Punta Rassa, and caused $258,000 in damage in Lee County. In September 1926 a hurricane with winds up to 140 mph produced 12-foot tides at Sanibel, Captiva, and Punta Rassa and 11-foot tides near Fort Myers. It caused about $3 million in damages from flooding. In October 1944 a hurricane with winds of 115 mph produced 8- to 11-foot tides in Lee County. In September 1947 a storm with sustained winds of 100 mph produced heavy rains, causing $2.0 million in flood damage in the Fort Myers area. In September 1960 Hurricane Donna was the most severe of the storms with 100-mph winds and 11-foot high tides in Lee County. *All of the barrier islands were overtopped with waters.* Property damage in Lee County alone was $16.5 million. The implications for future losses during hurricanes are enormous because of low elevations, exposed beaches, and the intensive development of the past decade. As of August 31, 1981, Lee County had 33,608 flood insurance policies valued at more than $1.347

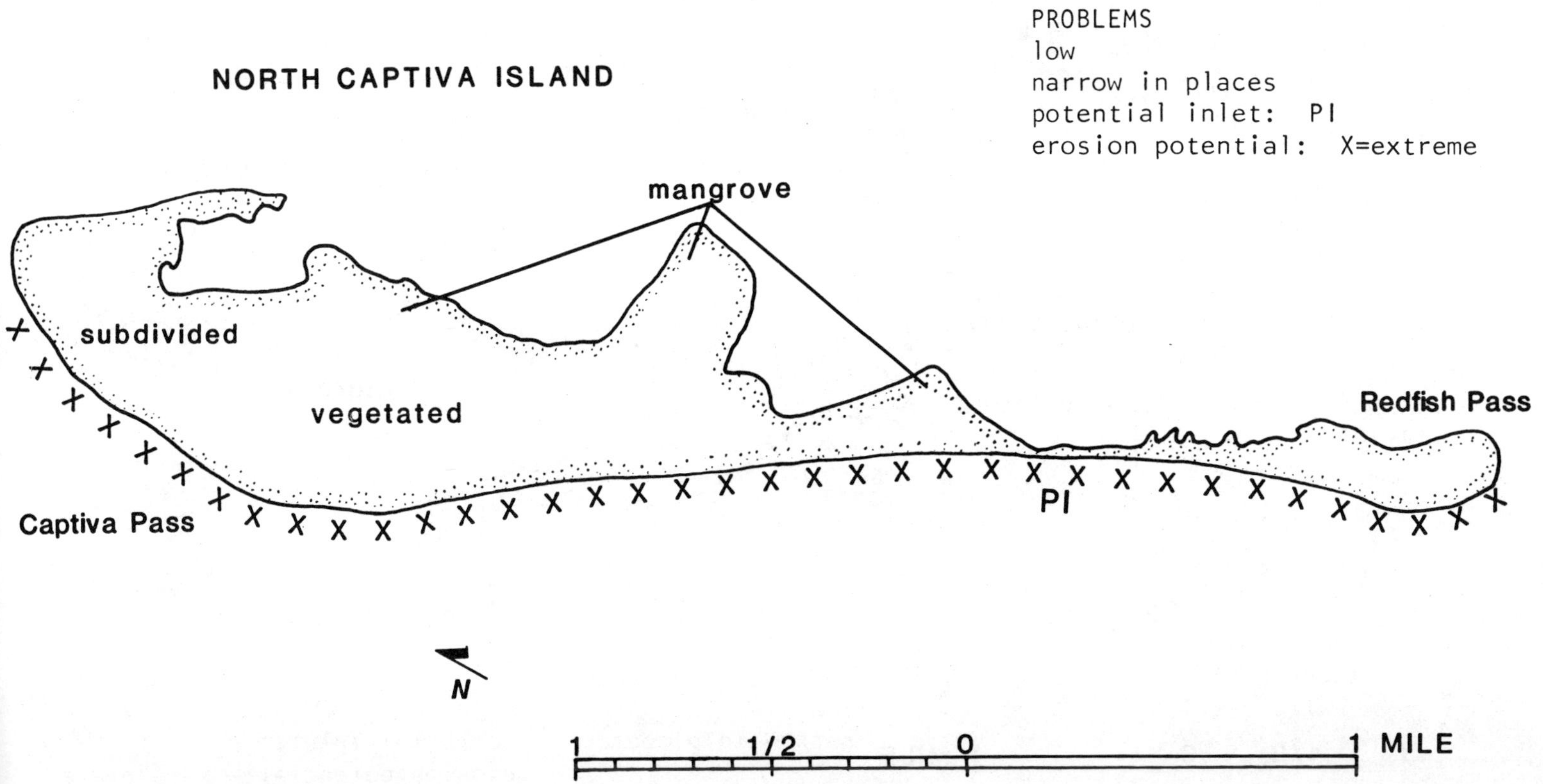

NORTH CAPTIVA ISLAND

PROBLEMS
low
narrow in places
potential inlet: PI
erosion potential: X=extreme

mangrove

subdivided

vegetated

Redfish Pass

Captiva Pass

PI

N

| 1 | 1/2 | 0 | 1 MILE |

Fig. 5.42. Site analysis map: North Captiva Island.

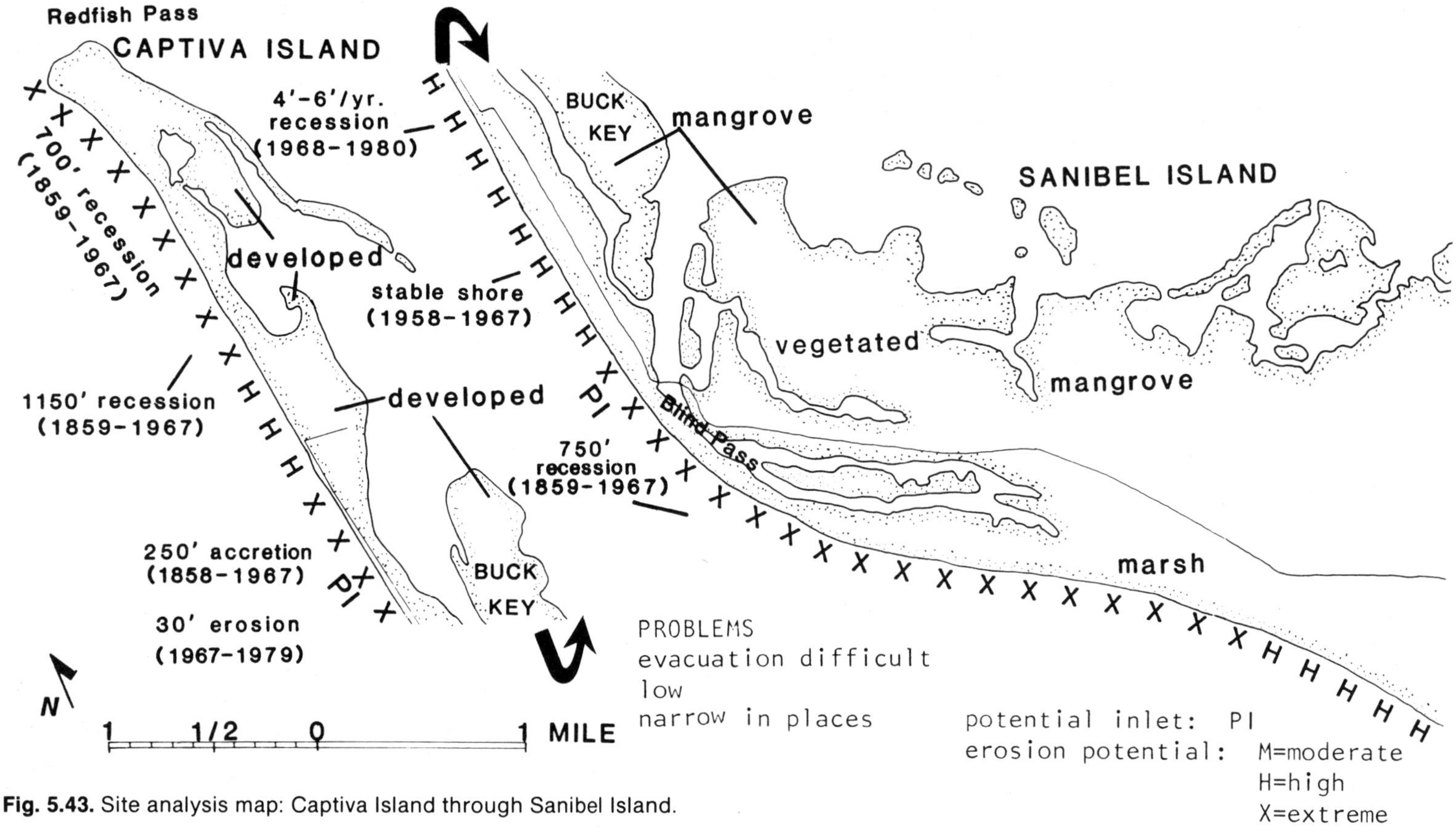

Redfish Pass
CAPTIVA ISLAND

4'-6'/yr.
recession
(1968-1980)

700' recession
(1859-1967)

developed

stable shore
(1958-1967)

1150' recession
(1859-1967)

developed

250' accretion
(1858-1967)

30' erosion
(1967-1979)

BUCK KEY

mangrove

SANIBEL ISLAND

vegetated

mangrove

Blind Pass

750'
recession
(1859-1967)

BUCK
KEY

marsh

N

1   1/2   0   1 MILE

PROBLEMS
evacuation difficult
low
narrow in places

potential inlet: PI
erosion potential: M=moderate
                    H=high
                    X=extreme

Fig. 5.43. Site analysis map: Captiva Island through Sanibel Island.

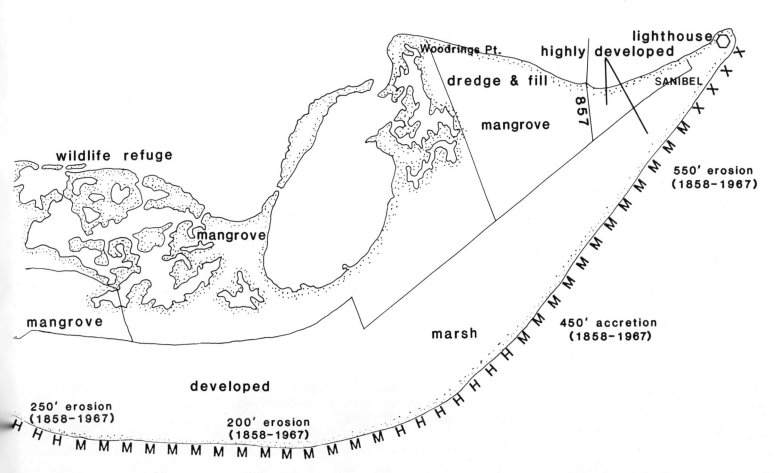

lighthouse

highly developed

Woodrings Pt.

dredge & fill

857

SANIBEL

mangrove

550' erosion
(1858-1967)

wildlife refuge

mangrove

450' accretion
(1858-1967)

mangrove

marsh

developed

250' erosion
(1858-1967)

200' erosion
(1858-1967)

billion. Unincorporated areas of Lee County's barrier islands are under the Emergency Flood Insurance Program, which provides only partial insurance coverage.

Gasparilla Island is a narrow island about 6.7 miles long, most of which is only between 5 to 7 feet above mean sea level (fig. 5.41). There is only 1 exit to the mainland at the northern end of the island. With the exception of a small area on the north end in Charlotte County, shoreline development is almost continuous. Old charts suggest that the island built out to the north about 750 feet in the last 100 years and that the southern end lost about 900 feet in the same period. The central section of the island has moved about 150 feet landward as well. Boca Grande Pass on the south is periodically dredged by the Corps of Engineers. At least 2 sheet-metal groins, 5 timber groins, 13 permeable concrete groins, 22 stone groins, and 9,600 feet of seawall have already been constructed on Gasparilla Island. About 70 feet of seawall on the southern end, part of a 700-foot section constructed by the county to protect the road, failed during Hurricane Gladys in October 1968. It was rebuilt in the same location.

A 200-foot terminal groin was built at the Boca Grande Historic Lighthouse Beach in 1970. Large amounts of sand were used to replenish this beach in 1971, 1977, and 1980. However, the beach and lighthouse area continue to erode. In a 1981 emergency, sandbags were used to protect the lighthouse. Recently another bigger, "better," $300,000, 300-foot terminal groin has been proposed to be placed 100 feet north of the last *terminal* groin.

Cayo Costa (or Lacosta; fig. 5.41) Island and North Captiva Island (fig. 5.42) are 7.3 and 4.3 miles long, respectively. Cayo Costa varies in width from about 300 feet at the narrows near the south end to 6,400 feet at its widest point just north of the middle of the island. The width of North Captiva varies from about 200 feet near the south end to about 3,200 feet near the middle. In general, these 2 islands have elevations of only 5 to 6 feet above mean sea level. Cayo Costa has about 10 structures, while North Captiva Island has over 40 existing structures and is undergoing slow development. Neither island is attached to the mainland. The northern end of the island near Boca Grande Pass and the southern end near Captiva Pass show steep scarps, indicating migration. The southern end of Cayo Costa is separated from the wider section by an old pass that has been closed off. Both islands are in a relatively natural state compared with their neighbors, and study of old charts shows that they are undergoing normal migration. Captiva Pass and Redfish Pass at the south end of North Captiva Island have not been dredged to this time. Captiva Island is about 5 miles long and varies somewhat from less than 200 feet wide at the south end to about 2,000 feet wide near the center (fig. 5.43). The narrow southern strip of sand represents the recent natural closure of Blind Pass, which used to separate Captiva Island from Sanibel Island. Most of the Gulf side is developed as residential and resort areas. The island is engineered with seawalls, groins, and beach renourishment and is experiencing serious ongoing erosion problems (fig. 5.44).

In 1961 the Captiva Erosion Prevention District installed 134 permeable groins on Captiva Island at a cost of $280,000 and an

way north of the timber groins. Despite these efforts, Hurricane Gladys and a number of tropical storms have repeatedly damaged the roadway that has then been rebuilt and repaired. At Turner Beach County Park at the southern end of the island more than $20,000 has been spent in erosion control and for relocation of public bathrooms. Along the southern 3.3 miles of beach there is a hodgepodge of seawalls, bulkheads, and revetments that may have temporarily protected the less than 100 mostly private dwellings behind them. However, with the concomitant loss of beach in front, there is a proposal to renourish this stretch of the island with 1.7 million cubic yards of sand, at an initial cost of $9 million, with follow-up maintenance and renourishment costs estimated at more than 1 million per year. In 1981 the northern 1.8 miles of Captiva Island (figure 5.43) was renourished with 715,000 cubic yards of sand at a cost of $2.8 million paid for by the resort property owners of South Sea Plantation.

Captiva Island is not connected directly to the mainland, but there is indirect access through Sanibel Island (fig. 5.43). As the history of road maintenance projects shows, the access road is subject to flooding in even modest storms.

Sanibel Island is among the most unusual and largest of the barrier islands on the Gulf coast. It has an approximate area of 11,675 acres, more than half of which is wetlands. The island is about 12 miles in length along the Gulf and varies in width from 700 feet to 13,000 feet, making it one of the widest barrier islands. However, elevations are generally below 6 feet on most of the island—much of it below 5 feet. With the exception of the northern

Fig. 5.44. Erosion at the north end of Captiva Island. The rocks placed here will, at best, only temporarily slow down erosion. Photo by Dinesh Sharma.

annual maintenance cost of $5,000 to $10,000. Subsequently, 2 timber groins were built in the middle of the island in 1966 at a cost of $20,000. In the mid-1960s the district and Lee County emplaced hundreds of thousands of cubic yards of fill at a cost of over $100,000 to protect the coastal highway. The county also spent over $12,000 to build rock revetments to protect the high-

2.7-mile section, the Gulf front is almost completely developed. The island had multiple curved dune ridges, most of which have been leveled for urban development in the past 10 to 20 years.

The northwesterly-trending portion of Sanibel's shoreline has been extremely dynamic due to the influence of Blind Pass. From 1858 to 1943 the northern half grew seaward about 320 feet. Since then the stability of the island's shoreline along the Gulf as well as on the back bays has been variable. The offshore depth contours for 6, 12, and 18 feet advanced seaward about 380, 370, and 650 feet, respectively, from 1880 to 1967 for the southern half of the island. Along the northern end of the island these depth contours retreated landward 150, 420, and 1,250 feet from 1956 to 1967.

On Sanibel Island only a half-dozen groins and a few thousand feet of seawalls have been constructed along the middle Gulf shore by individual property owners. The erosion problem appears to be more acute on the bay side of the island, and extensive seawalls and bulkheads have been built from the lighthouse to Wooderings Point. The only connection to the mainland is at the southeastern tip of the island.

Bunch Beach is a sheltered barrier beach welded to the mainland. From Sanibel Causeway to Pelican Bay the beach is about 4 miles long. It is mostly mangrove swamp with narrow beach berms. It is bordered on the mainland side by a freshwater slough known as Grassy Pond Slough. The width of the berm varies from 50 feet to about 150 feet, has some beach grasses, and is backed by red and black mangroves. In some places mangroves are exposed on the Gulf coast. There is little development on this beach.

Estero Island is also known as Fort Myers Beach (fig. 5.45). The Gulf shoreline is about 7 miles long, and the island width varies from 400 to 3,800 feet. The elevations are generally below 7 feet. In several locations along the island artificial canals from the bay side have been dug toward the beach, making the island vulnerable to breaching during hurricanes and tropical storms. Much of the Gulf front has been intensively urbanized for commercial, resort, and residential uses. Only the northern few acres at Bodwitch Point remain undeveloped. A new, ephemeral sand bar, locally called Little Estero Island, has been emerging along the southern 2.5 miles of the island. This sand bar is highly dynamic, unstable, and low in elevation. However, its presence provides some protection to the naturally eroding beaches on the southern end of Estero Island. Estero Island's Gulf shoreline is armored with 87 stone groins, 4 timber groins, 2 combination stone-timber groins, and more than 6,000 feet of linear concrete seawalls and some rock revetments. Additional seawalls, bulkheads, and revetments have been placed on the back bays of the island coastline. During the Matanzas Pass channel maintenance dredging in 1980, about 130,000 cubic yards of sand were placed along the northern 1.0 mile of Estero Island beach. The north end of Estero Island retreated 80 to 150 feet between 1957 and 1969 and up to about 400 feet on the southern end over the past 100 years.

There is 1 road link between Estero Island and the mainland located near the northwest end of the island. The southeast end is connected by road to Black Island.

The Lovers Key group of barrier islands consists of Black Island, Inner Key, and Lovers Key (fig. 5.46). The beach length is 2.7

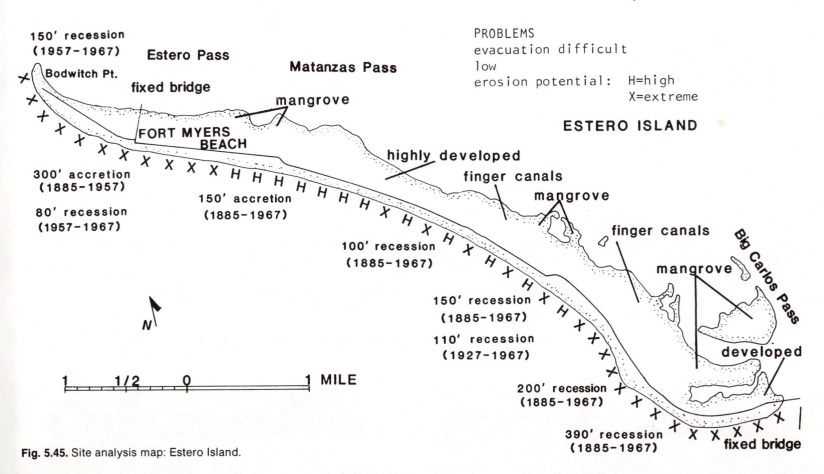

PROBLEMS
evacuation difficult
low
erosion potential: H=high
                      X=extreme

150' recession
(1957-1967)

Bodwitch Pt.

Estero Pass

fixed bridge

Matanzas Pass

mangrove

FORT MYERS
BEACH

ESTERO ISLAND

highly developed

finger canals

mangrove

finger canals

300' accretion
(1885-1957)

80' recession
(1957-1967)

150' accretion
(1885-1967)

mangrove

Big Carlos Pass

100' recession
(1885-1967)

150' recession
(1885-1967)

110' recession
(1927-1967)

developed

200' recession
(1885-1967)

390' recession
(1885-1967)

fixed bridge

N

1   1/2   0   1 MILE

Fig. 5.45. Site analysis map: Estero Island.

miles, and the total area for the Lovers Key group is about 300 acres of upland and wetlands. The width of Lovers Key on the Gulf varies from 30 feet to about 900 feet. Inner Key, which used to be on the Gulf prior to the formation in 1952 of the present-day Lovers Key, is about 600 feet at the widest point, including mangroves. Elevations are generally below 5 feet on Black Island, 3 feet on Inner Key, and 2 to 4 feet on Lovers Key. There is no urban development on these islands at this time, with the exception of State Highway 865 and county park facilities.

The Lovers Key group of islands has no erosion control structures along the beach, except for 500 feet of rock revetment that have been placed on the northwest tip of Black Island.

Little Hickory Island, also called Bonita Beach Island (fig. 5.46), is about 6 miles long and varies from 300 feet to 3,200 feet in width. Only the northern half of the island is in Lee County. The remaining half is located in Collier County. The natural berm is about 5 feet high. The entire Gulf front in Lee County, with the exception of the southern 250 feet and northern 1.25 miles, has been intensively developed for resort and residential use. Bonita Beach Island has about 3,000 feet of linear seawall on the northern 1 mile. A permit to place rock revetments along 600 feet of the Bonita Beach Club seawall has been approved recently by the Florida Department of Natural Resources, in spite of findings that the revetment would increase the erosion along the beach, steepen the offshore depth contours, and possibly restrict public access to the beaches on the north.

Collier County (figures 5.46–5.50)

Collier County has 50 miles of Gulf shoreline, of which 35 miles are sandy beaches and 15 miles south of Cape Romano Island are generally mangrove islands with small, localized beaches. The northern 20 miles of sandy beaches consist of Bonita Beach Island bounded by Wiggins Pass. The remaining beach is part of the mainland cut by Clam Pass, Doctors Pass, and Gordon Pass. Islands behind the Gulf beach are made up of material (known as spoil) removed from navigational channels. They now act in a manner similar to barrier islands. The remaining 15 miles south of Gordon Pass consist of numerous barrier islands: Keewaydin Island bounded by Little Marco Pass, the little Marco Island group bounded by Hurricane Pass, several unnamed islands bounded by Big Marco Pass, Marco Island bounded by Caxambas Pass, Kice Island bounded by Blind Pass (presently closed), Morgan Beach bounded by Morgan Pass, and Cape Romano Island bounded by Gullican Bay.

The barrier beaches of Collier County are composed of fine quartz sand shell fragments in varying proportions. Almost all the beaches and islands are between 5 to 8 feet high (norm about 6 feet), with the exception of small areas on Naples City Beach and Marco Island that are above 10 feet. There are no pronounced dunes.

Bonita Beach, or Little Hickory Island (the northern 3 miles in Collier County), are low and vary in width from about 2,500 feet in the north to 300 feet near the middle portion (fig. 5.46). Wiggins

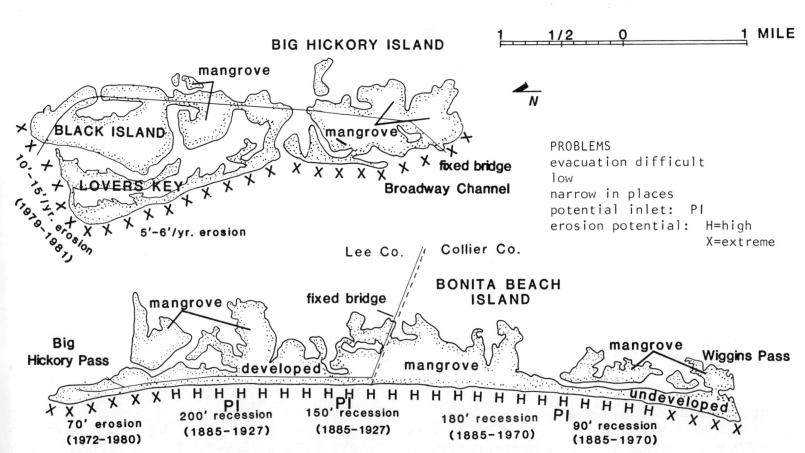

Fig. 5.46. Site analysis map: Lovers Key through Bonita Beach Island.

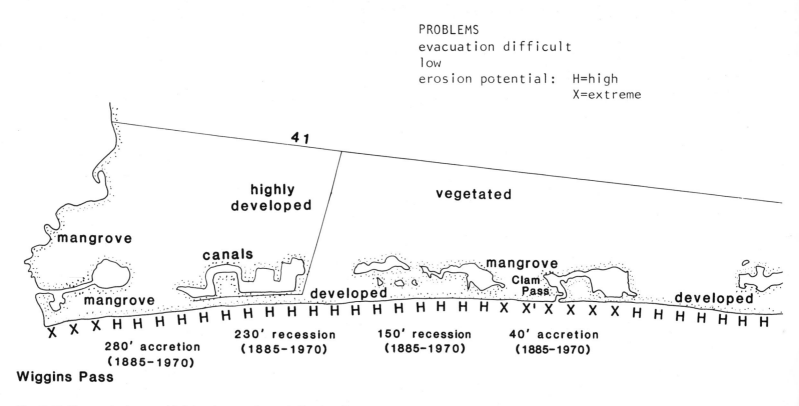

PROBLEMS
evacuation difficult
low
erosion potential: H=high
                     X=extreme

41

highly
developed

vegetated

mangrove

canals

mangrove

mangrove

Clam
Pass

developed

developed

mangrove

X X X H H H H H H H H H H H H H H H H X X' X X X X X H H H H H H
X

280' accretion
(1885-1970)

230' recession
(1885-1970)

150' recession
(1885-1970)

40' accretion
(1885-1970)

Wiggins Pass

Fig. 5.47. Site analysis map: Mainland coast through Gordon Pass.

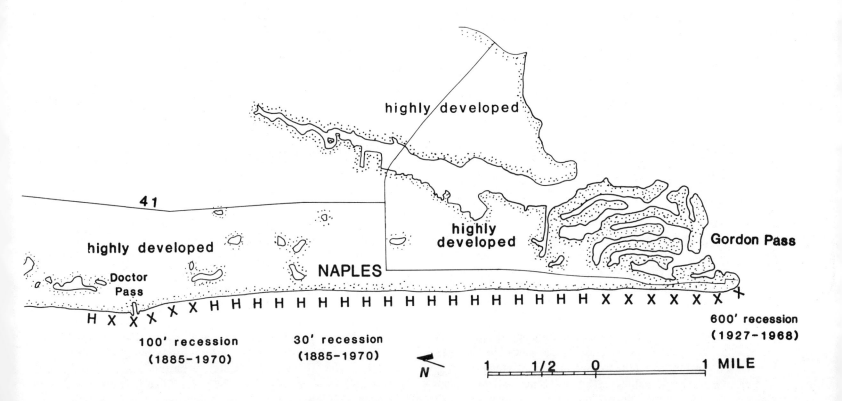

highly developed

41

highly developed

highly
developed

Gordon Pass

Doctor
Pass

NAPLES

H X X X X X H H H H H H H H H H H H H H H H H H X X X X X X +

100' recession
(1885-1970)

30' recession
(1885-1970)

600' recession
(1927-1968)

N

1   1/2   0   1 MILE

Pass is an unstabilized inlet with a channel depth of about 6 feet. Total area of Bonita Beach or Little Hickory Island in Lee and Collier counties is about 1,428 acres, of which 1,235 acres are wetlands.

The Wiggins Pass to Clam Pass barrier island shoreline is about 4.7 miles long and covers about 1,240 acres, of which about 844 acres are wetlands. The island is 300 feet to 2,500 feet wide. The elevations of the island dip from a high of 5 feet along the beach toward the back bay wetlands. Clam Pass is a natural unimproved pass 3 to 4 feet deep. The northern 1 mile of beach is owned by the state and maintained as Del-Nor State Recreation Area.

The man-made barrier island from Clam Pass to Doctors Pass is about 3 miles long and varies in width from 250 feet in the north to 1,800 feet in some parts in the middle. It is bordered by Inner Doctors Bay and has an approximate area of 245 acres, of which about half is wetlands. The island has a 5-foot-high berm along much of its length, except near the southern end of Outer Clam Bay. Doctors Pass was stabilized by the city of Naples with 2 short stone jetties. The pass varies in depth from 4 to 10 feet and is 150 feet wide.

Naples City Beach is 5.6 miles long, bounded on the north by Doctors Pass and on the south by Gordon Pass (fig. 5.47), and has been stabilized with timber-stone terminal groins to provide a navigation channel 12 feet deep and 150 feet wide. The Gulf beach is generally low and only 20 to 50 feet wide. Extensive dredge-and-fill finger canal development on the south end near Naples Bay make this area vulnerable to breaching during storms.

The Keewaydin Island group is about 9.3 miles long and 2,980 acres in area (fig. 5.48). There are 2,230 acres of wetland and 750 acres of upland in the unit. The island is bounded on the south by Little Marco Pass. Elevations are generally 3 to 6 feet along the beach and dip toward the back bay. The width varies from about 4,000 feet in the north to 200 feet in the south. The beaches are generally narrow and steep with erosional scarps in several places.

Little Marco Island is about 1.5 miles long and bounded by Hurricane Pass and Little Marco Pass. Much of the island is below 5 feet in elevation and less than 1,500 feet wide. This island used to be exposed to the Gulf, but the southerly buildup of Keewaydin Island now protects it. Hurricane Pass is a natural pass with depths of 8 feet. Big Marco Pass is a natural pass with depths up to 30 feet.

Marco Island is about 4.6 miles long and 5,100 acres in size (fig. 5.49). Approximately 785 acres of wetlands remain on the island after extensive dredging and filling in the 1970s. Much of the island is 4 to 7 feet in elevation, with 1 small area north of the Marriott Hotel exceeding 10 feet. Marco Island has beaches from 100 to 150 feet in width. The northern 0.5 mile of beach front is owned by Collier County.

Kice Island, Morgan Beach, and Cape Romano appear to be a unified barrier island group (fig. 5.50). These islands have about 4.8 miles of sandy beach with a total area of about 2,650 acres (much of which is wetlands) behind the narrow beach berm. There are 3 passes, Caxambas Pass in the north, Blind Pass in the middle (now closed), and Morgan Pass in the south. The Cape Romano

Island group is the southernmost sandy barrier island along the eastern Gulf of Mexico. The beaches along much of this island group are quite narrow, steep, unstable, low, and subject to frequent overwash and inlet formation. The entire island group is below 5 feet in elevation—much of it less than 2 to 3 feet. Approximately 958 acres of this island are owned by the state of Florida.

During Hurricane Donna in 1960, tides of 11 feet were reported along Naples beaches. The 100-year flood heights along Collier County are 12.5 to 14.0 feet above mean sea level.

Tropical storms and hurricanes, and their attendant waves and currents, reshape the beaches periodically. Since 1900 a total of 41 hurricanes passed within 150 miles of Naples, or an average of 1 hurricane every 2 years. Tropical storms and winter storms are most frequent and cause changes and damage to the beaches as well. Some of the notable hurricanes that severely affected Collier County were the storm of 1873 that produced 14-foot tides at Punta Rassa in Lee County and 90-mph winds; the October 1910 hurricane that featured 70- to 90- mph winds and 10.5-feet tides in Everglades; the October 1921 hurricane that created tides 1 to 11 feet above normal in Collier County; the Labor Day hurricane of 1926 that brought winds of 90 to 115 mph and tides of more than 6 feet at Everglades. The October 1944 hurricane produced 7.4 foot tides at Everglades and 8 to 11 foot tides along Naples Beach; flooding depths of 5 feet were reported along the low areas of Naples and Everglades. In the 1944 storm considerable damage occurred along the beach, severe erosion took place, and more than 4 miles of wooden sheet-pile bulkhead were destroyed, along with a 1,000-foot pier. The hurricane of September 1947 produced tides 5.5 feet above normal, flooding many low areas in Naples and Everglades with water 2 to 3 feet deep. Hurricane Donna of September 1960 was the most severe storm to affect this area in this century. A high tide of 9.0 feet inundated Everglades and neighboring beaches, and flooding extended 6 to 10 miles inland. All the barrier beaches were overtopped and flooded. Low areas in Naples were flooded 3 to 4 feet deep. Tides and heavy wave action eroded beach berms and covered the beach roads with several feet of sand. Street ends were heavily eroded and undermined. Nearly all homes on Vanderbilt Beach and other exposed beaches were severely damaged or destroyed. Nearly 300 homes and trailers in Collier County suffered heavy damage or were completely lost. Almost all of the seawalls, bulkheads, and groins suffered major damage. The Naples city piers were again destroyed. The total flood damage in Collier County exceeded $8.5 million (at 1960 prices). Hurricane Betsy in September 1965 created 5-foot tides at Everglades and cost $166,000 in storm damage in Collier County. As of August 1981, more than 13,800 federal flood insurance policies valued at over $785 million had been sold along the Collier County coast. If a major hurricane strikes this coast in the 1980s, property damage could well reach several billion dollars.

The shoreline of Collier County has a history of advance and erosion from 1885 to 1970. From the northern county line to Wiggins Pass, there was a general trend toward erosion from 1885 to 1970. At the north end of Wiggins Pass, the shoreline retreated 190 feet form 1885 to 1970. The south end of the pass advanced

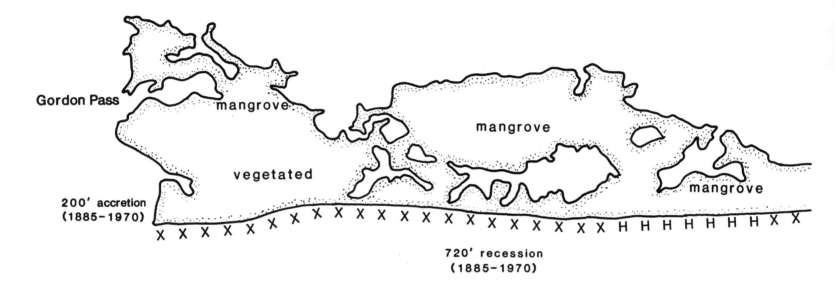

Gordon Pass

mangrove

vegetated

mangrove

mangrove

200' accretion
(1885–1970)

720' recession
(1885–1970)

Fig. 5.48. Site analysis map: Keewaydin Island.

280 feet during this period. The entire beach from Wiggins Pass to Clam Pass eroded 50 to 230 feet from 1885 to 1970.

The beach between Clam Pass and Doctors Pass advanced and retreated. Between 1885 and 1928 the beach eroded about 50 feet, and from 1927 to 1970 it advanced 55 feet. The north end of Doctors Pass retreated 100 feet from 1885 to 1970.

The shoreline from Doctors Pass to Gordon Pass also both advanced and eroded from 1885 to 1980. From 1885 to 1927 the shoreline advanced an average of 170 feet; from 1927 to 1968 it receded an average of 220 feet; and from 1970 to 1980 the shoreline advanced again an average of 17 feet, the net result being an average retreat of about 45 feet.

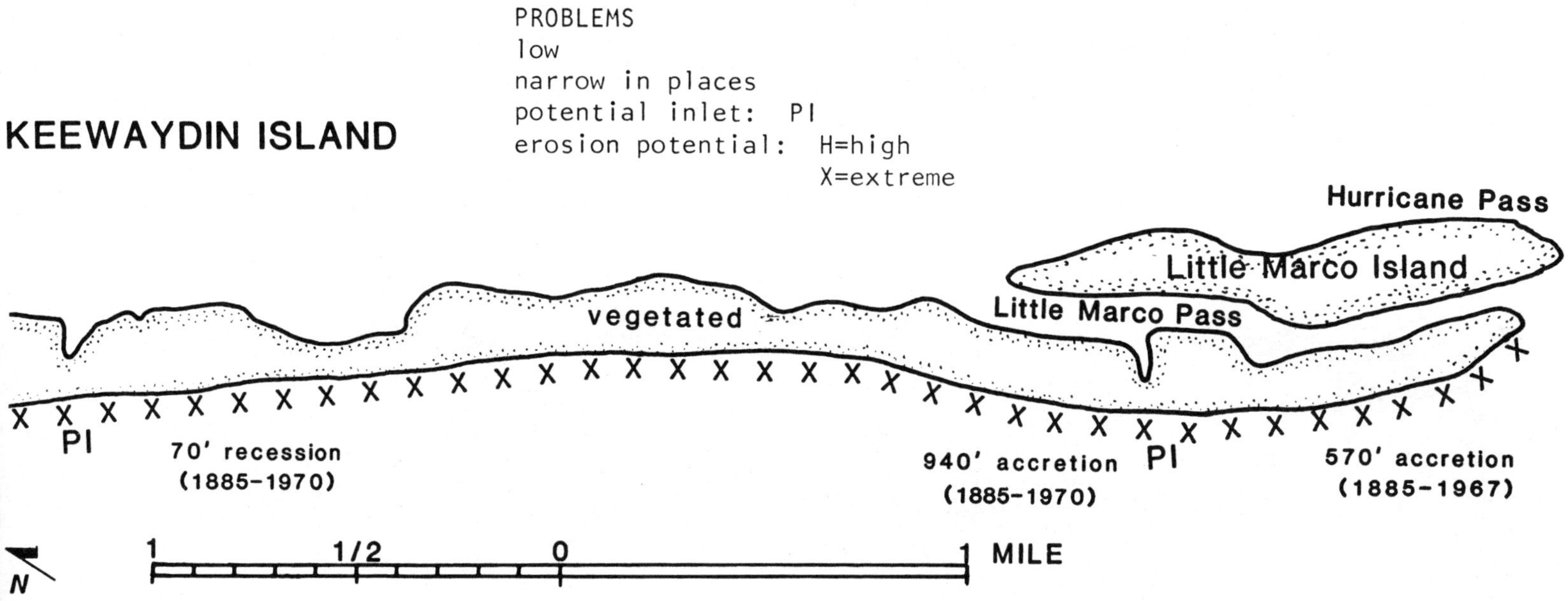

KEEWAYDIN ISLAND

```
PROBLEMS
low
narrow in places
potential inlet: PI
erosion potential: H=high
                   X=extreme
```

Hurricane Pass

Little Marco Island

Little Marco Pass

vegetated

PI

70' recession
(1885-1970)

940' accretion
(1885-1970)

PI

570' accretion
(1885-1967)

1   1/2   0   1 MILE

N

Keewaydin Island has a history of advance and retreat and has several historic inlets, potential inlets, and overwash areas. From 1885 to 1970 the shoreline advanced an average of 150 feet, with the greatest advance being south of Gordon Pass (200 feet) and north of Little Marco Pass (570 feet). Since 1970 the southern end of the island has built out several thousand feet in the southerly direction and the inlet has migrated south. The area 1.5 miles south of Gordon Pass retreated more than 720 feet between 1885 and 1970 and shows steep erosional scarps and falling trees.

The shoreline between Little Marco Pass and Big Marco Pass has retreated during the entire period from 1885 to 1970, averaging about 640 feet. The shifting of the Little Marco Pass channel

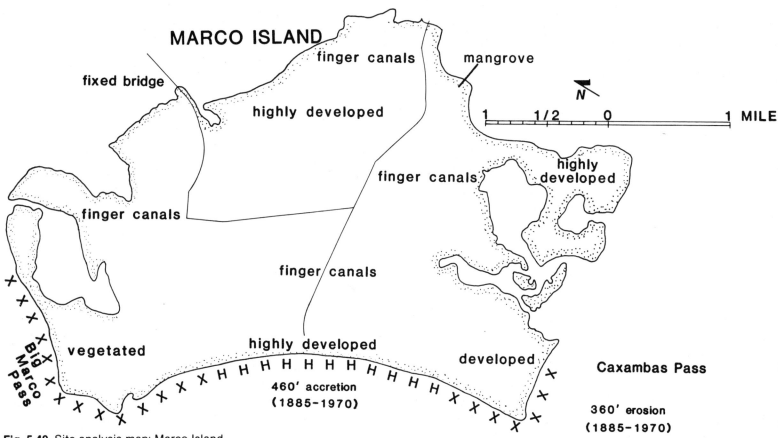

Fig. 5.49. Site analysis map: Marco Island.

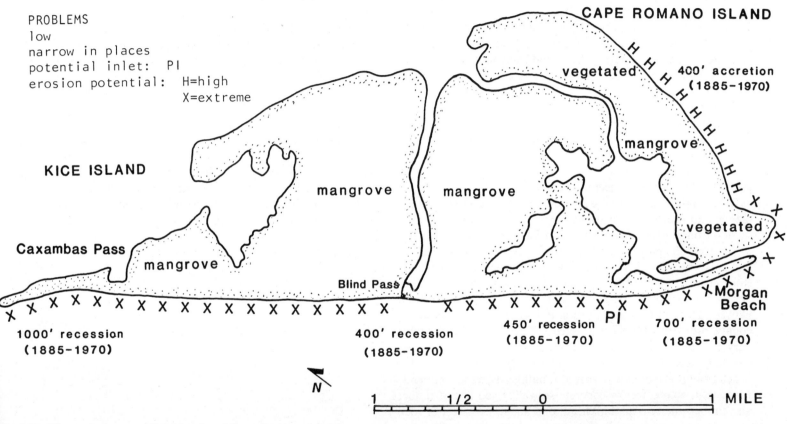

PROBLEMS
low
narrow in places
potential inlet: PI
erosion potential: H=high
                    X=extreme

CAPE ROMANO ISLAND

vegetated   400' accretion
                 (1885-1970)

mangrove

KICE ISLAND

mangrove   mangrove

vegetated

Caxambas Pass

mangrove

Blind Pass   Morgan
                                          Beach

1000' recession   400' recession   450' recession   PI   700' recession
(1885-1970)   (1885-1970)   (1885-1970)   (1885-1970)

N

1   1/2   0   1 MILE

Fig. 5.50. Site analysis map: Kice Island and Cape Romano Island.

(as well as Hurricane Pass channel) has caused extensive erosion and loss of beaches on Little Marco Island and 2 other unnamed islands. The mean high water shoreline on Marco Island advanced an average of 216 feet during the period from 1885 to 1970, with the exception of the southern 0.5 mile, where it retreated an average of 360 feet.

Kice Island and Cape Romano experienced about 220 to 1,100 feet of retreat along the entire stretch of the beach from 1885 to 1970. Most severe retreat was near Caxambas Pass, Blind Pass (presently closed), and Morgan Pass (which has migrated south more than 1.25 miles since 1885). There are several old inlet sites, and the entire area is subject to overwash.

In the past 30 years groins, jetties, seawalls, bulkheads, revetments, and beach fills of various types have been built along the section from Doctors Pass to Gordon Pass and on Marco Island to "protect" the property and "control" erosion. In the late 1950s and 1960s, the city of Naples and property owners constructed 37 stone groins, 26 timber groins, 4 concrete H-pile groins with timber panels, 4 timber-pile groins, and a concrete dog-bone groin between Doctors Pass and Gordon Pass. In the remainder of the county, property owners constructed 5 stone groins. These groins vary greatly in length, but they average about 150 feet. Between Gordon Pass and Doctors Pass more than 72 percent of the beach front is armored with seawalls, bulkheads, and revetments. More than 20,000 feet of concrete seawall, 3,500 feet of timber and timber-pile seawall/bulkhead, and about 3,000 feet of rock revetments have been built. In 1959 and 1960 Doctors Pass was straightened and 2 jetties constructed. In mid-1966, during the maintenance dredging, an undetermined amount of spoil was placed on the south side of Doctors Pass. According to the U.S. Army Corps of Engineers, "this material did not remain in place very long." In 1962 the initial dredging of Gordon Pass was authorized. Since then the pass has been dredged 3 times. In 1968, 8,800 cubic yards of spoil were removed; in 1970, 181,400 cubic yards; and in 1980, 235,000 cubic yards. Total cost amounted to $1,357,500. The spoil from this dredging was placed on Keewaydin Island south of the pass, which temporarily helped to build the northern beaches on Keewaydin. Maintenance dredging of 23,000 cubic yards annually has been required. Some of this sand is coming from the beaches north of the pass. Private developers have placed undetermined amounts of sand on various parts of Marco Island.

6. Building or buying a house near the beach

In reading this book you may conclude that the authors seem to be at cross-purposes. On the one hand, we point out that building on the coast is risky. On the other hand, we provide you with a guide to evaluate the risks, and in this chapter we describe how best to buy or build a house near the beach.

This apparent contradiction is more rational than it might seem at first. For those who will heed the warning, we describe the risks of owning shorefront property. But we realize that coastal development will continue as some individuals will always be willing to gamble with their fortunes to be near the shore. For those who elect to play this game of real estate roulette, we provide some advice on improving the odds, on reducing (not eliminating) the risks. We do not recommend, however, that you play the game!

If you want to learn more about construction near the beach, we recommend the book *Coastal Design: A Guide for Builders, Planners, and Home Owners* (New York: Van Nostrand Reinhold, 1983), which gives more detail on coastal construction and may be used to supplement this volume. In addition, the Federal Emergency Management Agency's *Design and Construction Manual for Residential Buildings in Coastal High Hazard Areas* is an informative manual for coastal construction that contains much reference material. Also the Florida Department of Natural Resources' *Coastal Construction Building Code Guidelines* when combined with the South Florida or Standard (Southern Standard) Building Code make outstanding guides for the construction of residential and commercial structures. The DNR publication also has a comprehensive bibliography.

Coastal realty versus coastal reality

Coastal property is not the same as inland property. Do not approach it as if you were buying a lot in a developed part of the mainland or a subdivided farm field in the coastal plain. The previous chapters illustrate that the shores of Florida, especially the barrier islands, are composed of variable environments and are subjected to nature's most powerful and persistent forces. The reality of the coast is its dynamic character. Property lines are an artificial grid superimposed on this dynamism. If you choose to place yourself or others in this zone, prudence is in order.

A quick glance at the architecture of the structures on the Florida coast provides convincing evidence that the reality of coastal processes was rarely considered in their construction. Not too many years back, old-timers wisely lived behind the protection of sand dunes; only recently have city-bred, over-civilized people built in front of dunes to better see the storm come in. Except for meeting minimal building code requirements, no further thought seems to have been given to the safety of many of these buildings. The failure to follow a few basic architectural guidelines that recognize this reality will have disastrous results in the next major storm.

Life's important decisions are based on an evaluation of the facts. Few of us buy goods, choose a career, take legal, financial, or medical actions without first evaluating the facts and seeking advice. In the case of coastal property, two general areas should be considered: site safety and the integrity of the structure relative to the forces to which it will be subjected.

A guide to evaluating the site(s) of your interest on the West Florida Gulf shoreline is presented in chapter 5, along with hazard evaluation maps.

The structure: concept of balanced risk

A certain chance of failure for any structure exists within the constraints of economy and environment. The objective of building design is to create a structure that is both economically feasible and functionally reliable. A house must be affordable and have a reasonable life expectancy free of being damaged, destroyed, or wearing out. In order to obtain such a house, a balance must be achieved among financial, structural, environmental, and other special conditions. Most of these conditions are heightened on the coast—property values are higher, there is a greater desire for aesthetics, the environment is more sensitive, the likelihood of storms is greater, and there are more hazards with which to deal.

The individual who builds or buys a home in an exposed area should fully comprehend the risks involved and the chance of harm to home or family. The risks should then be weighed against the benefits to be derived from the residence. Similarly, the developer who is putting up a motel should weigh the possibility of destruction and death during a hurricane versus the money and other advantages to be gained from such a building. Then and only then should construction proceed. For both the homeowner and the developer, proper construction and location reduce the risks involved.

The concept of balanced risk should take into account the following fundamental considerations:

1. A coastal structure, exposed to high winds, waves, or flooding, should be stronger than a structure built inland.
2. A building with high occupancy, such as an apartment building, should be safer than a building with low occupancy, such as a single-family dwelling.
3. A building that houses elderly or sick people should be safer than a building housing able-bodied people.
4. Because construction must be economically feasible, ultimate and total safety is not obtainable for most homeowners on the coast.
5. A building with a planned long life, such as a year-round residence, should be stronger than a building with a planned short life, such as a mobile home.

Structures can be designed and built to resist all but the largest storms and still be within reasonable economic limits.

Structural engineering is the designing and constructing of

buildings to withstand the forces of nature. It is based on a knowledge of the forces to which the structures will be subjected and an understanding of the strength of building materials. The effectiveness of structural engineering design was reflected in the aftermath of Typhoon Tracy that struck Darwin, Australia, in 1974: 70 percent of housing that was not based on structural engineering principles was destroyed and 20 percent was seriously damaged; only 10 percent of the housing weathered the storm. In contrast, more than 70 percent of the structurally engineered large commercial, government, and industrial buildings came through with little or no damage, and less than 5 percent of such structures were destroyed. Because housing accounts for more than half of the capital cost of the buildings in Queensland, the state government established a building code that requires standardized structural engineering for houses in hurricane-prone areas. This improvement has been achieved with little increase in construction and design costs.

Coastal forces: design requirements

Hurricanes, with their associated high winds and storm surge topped by large waves are the most destructive of the forces to be reckoned with on the coast. Figure 6.1 illustrates the effects of hurricane forces on houses and other structures.

Hurricane winds

Estimates of wind velocity to be used in designing structures along the Florida coast vary somewhat with the building codes. The South Florida Building Code specifies a wind velocity of 120 mph at a height of 30 feet above the ground. The Standard Building Code uses a map to indicate the maximum wind velocity for a given location. It shows winds from 110 to 130 mph along the Florida coast. (The city of Sanibel specifies 130 mph in its code.) Florida's *Coastal Construction Building Code Guidelines* states that "for habitable structures within the coastal construction building zone, the design wind velocity for load computations shall be a minimum of 140 m.p.h at a height of 30 feet above the ground."

The velocity of the wind can be evaluated in terms of the pressure exerted. The pressure varies with the square of the velocity, the height above the ground, and the shape of the object against which it is blowing. A 140-mph wind will exert a pressure about twice that of a 100-mph wind (as $[140/100]^2 = 1.96$). The above-mentioned *Guidelines* offers a table that we quote in part.

Height above ground (in feet)	Minimum velocity pressure (in pounds per square feet)
0 to 5	30
25 to 35	50
100 to 150	75
over 1000	135

The above pressures must be multiplied by shape factors to be obtained from either the South Florida Building Code or the Standard Building Code, whichever applies to your locality.

As an example, the Standard Building Code gives a shape factor of 1.4 for a flat vertical surface, such as a sign. This means that a

WIND

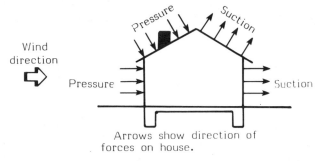

Wind direction

Arrows show direction of forces on house.

DROP IN BAROMETRIC PRESSURE

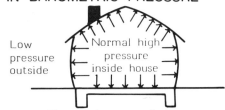

Low pressure outside

Normal high pressure inside house

The passing eye of the storm creates different pressure inside and out, and high pressure inside attempts to burst house open.

WAVES

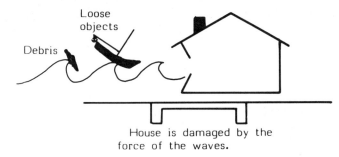

Loose objects

Debris

House is damaged by the force of the waves.

HIGH WATER

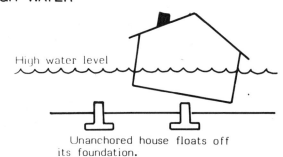

High water level

Unanchored house floats off its foundation.

Fig. 6.1. Forces to be reckoned with at the shoreline.

140-mph wind at 30 feet would exert a pressure of $1.4 \times 50 = 70$ psf against the flat surface instead of the basic 50 psf listed in the table. The effective pressure on a curved surface such as a sphere or a cylinder is less than on a flat surface.

Wind velocity increases with height above ground, so a tall structure is subject to greater velocity and thereby greater pressure than a low structure.

A house or building designed for inland areas is built primarily to resist vertical loads. It is assumed that the foundation and framing must support the load of the walls, floor, roof, and furniture with relatively insignificant wind forces.

A well-built house in a hurricane-prone area, however, must be constructed to withstand a variety of strong wind forces that may come from any direction. Although many people think that wind damage is caused by uniform horizontal pressures (lateral loads), most damage, in fact, is caused by uplift (vertical), suctional (pressure-outward from the house), and twisting (torsional) forces. High horizontal pressure on the windward side is accompanied by suction on the leeward side. The roof is subject to downward pressure and, more importantly, to uplift. Often a roof is sucked up by the uplift drag of the wind. Usually the failure of houses is in the devices that tie the parts of the structure together. All structural members (beams, rafters, columns) should be fastened together on the assumption that about 25 percent of the vertical load on the member may be a force coming from any direction (sideways or upwards). Such structural integrity is also important if it is likely that the structure may be moved to avoid destruction by shoreline retreat.

Storm surge

Storm surge is a rise in sea level above the normal water level during a storm. During hurricanes the coastal zone is inundated by storm surge and accompanying storm waves, and these cause most property damage and loss of life.

Often the pressure of the wind backs water into streams or estuaries already swollen from the exceptional rainfall brought on by the hurricane. Water is piled into the bays between islands and the mainland by the offshore storm. In some cases the direction of flooding may be from the bay side of the island. This flooding is particularly dangerous when the wind pressure keeps the tide from running out of inlets, so that the next normal high tide pushes the accumulated waters back and higher still.

Flooding can cause an unanchored house to float off its foundation and come to rest against another house, severely damaging both. Disaster preparedness officials have pointed out that it is a sad fact that even many condominiums built on pilings are not anchored or tied to those pilings, just set on top. Even if the house itself is left structurally intact, flooding may destroy its contents. People who have cleaned the mud and contents of a house subjected to flooding retain vivid memories of its effects.

Proper coastal development takes into account the expected level and frequency of storm surge for the area. In general, building standards require that the first habitable floor of the dwelling be above the 100-year flood level plus an allowance for wave height. At this level, a building has a 1 percent probability of being flooded in any given year.

Hurricane waves

Hurricane waves can cause severe damage not only in forcing water onshore to flood buildings but also in throwing boats, barges, piers, houses, and other floating debris inland against standing structures. The force of a wave may be understood when one considers that a cubic yard of water weighs over three-fourths of a ton: hence, a breaking wave moving shoreward at a speed of several tens of miles per hour can be one of the most destructive elements of a hurricane. Waves also can destroy coastal structures by scouring away the underlying sand, causing them to collapse. It is possible to design buildings to survive crashing storm surf as many lighthouses, for example, have survived this. But in the balanced-risk equation, it usually is not economically feasible to build ordinary cottages to resist the more powerful of such forces. On the other hand, cottages can be made considerably more storm-worthy by following the suggestions in the rest of this chapter.

Barometric pressure change

Changes in barometric pressure also may be a minor contributor to structural failure. If a house is sealed at a normal barometric pressure of 30 inches of mercury, and the external pressure suddenly drops to 26.61 inches of mercury (as it did in Hurricane Camille in Mississippi in 1969), the pressure exerted within the house would be 245 pounds per square foot. An ordinary house would explode if it were leakproof. In tornadoes, where there is a severe pressure differential, many houses do burst. In hurricanes the problem is much less severe. Fortunately, most houses leak, but they must leak fast enough to prevent damage. Given the most destructive forces of hurricane wind and waves, pressure differential may be of minor concern. Venting the underside of the roof at the eaves is a common means of equalizing internal and external pressure.

Figure 6.2 illustrates some of the actions that a homeowner can take to deal with the forces just described.

House selection

Some types of houses are better than others at the shore, and an awareness of the differences will help you make a better selection, whether you are building a new house or buying an existing one.

Worst of all are unreinforced houses, whether they are brick, concrete block, hollow clay-tile, or brick veneer because they cannot withstand the lateral forces of wind and wave and the settling of the foundation. Adequate and extraordinary reinforcing in coastal regions will alleviate the inherent weakness of unit masonry, if done properly. Reinforced concrete and steel frames are excellent but are rarely used in the construction of small, residential structures.

It is hard to beat a wood-frame house that is properly braced and anchored and has well-connected members. The well-built wood house will often hold together as a unit, even if moved off its foundations, when other types disintegrate. Although all of the structural types noted above are found in the coastal zone, newer structures tend to be of the elevated wood-frame type.

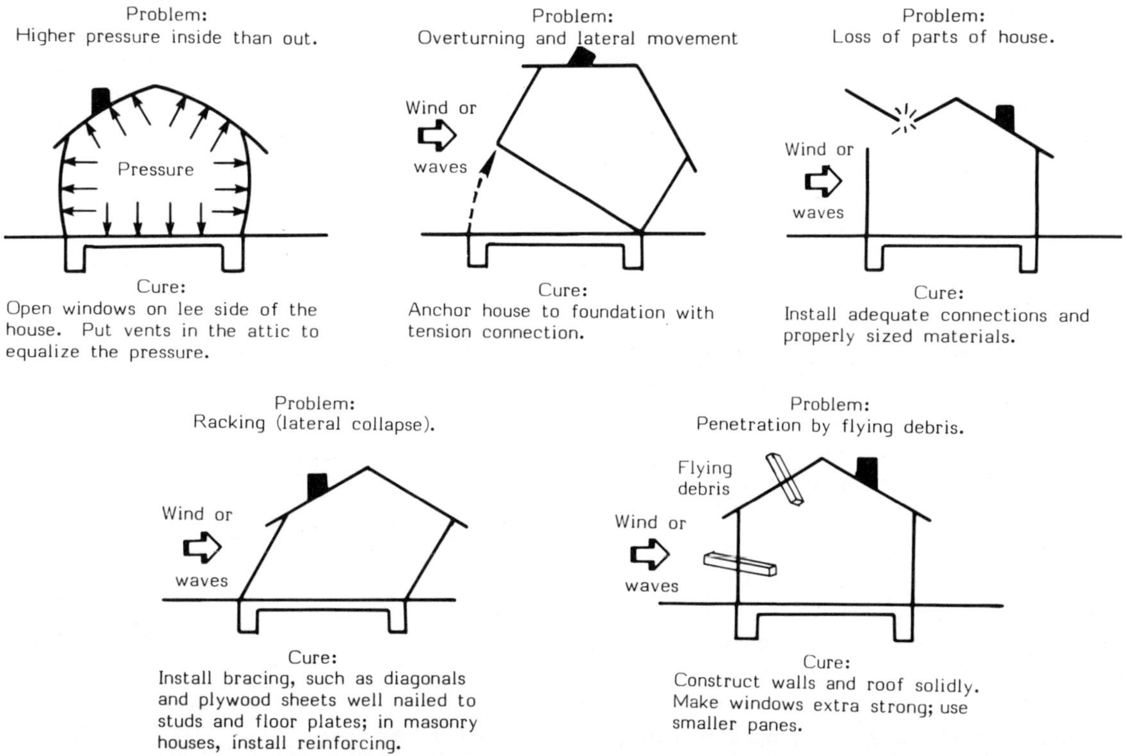

Problem:
Higher pressure inside than out.

Pressure

Cure:
Open windows on lee side of the
house. Put vents in the attic to
equalize the pressure.

Problem:
Overturning and lateral movement

Wind or
waves

Cure:
Anchor house to foundation with
tension connection.

Problem:
Loss of parts of house.

Wind or
waves

Cure:
Install adequate connections and
properly sized materials.

Problem:
Racking (lateral collapse).

Wind or
waves

Cure:
Install bracing, such as diagonals
and plywood sheets well nailed to
studs and floor plates; in masonry
houses, install reinforcing.

Problem:
Penetration by flying debris.

Flying
debris

Wind or
waves

Cure:
Construct walls and roof solidly.
Make windows extra strong; use
smaller panes.

Fig. 6.2. Modes of failure and how to deal with them. Modified from U.S. Civil Defense Preparedness Agency
Publication TR-83.

Keeping dry: pole or stilt houses

On coastal regions subject to flooding by waves or storm surge, the best and most common method of minimizing damage is to raise the lowest floor of a residence above the expected level. Also, the first habitable floor of a home must be above the 100-year storm-surge level (plus calculated wave height) to qualify for federal flood insurance. As a result, most modern flood-zone structures should be constructed on piling, well anchored in the sub-soil. Elevating the structure by building a mound is not suited to the coastal zone because mounded soil is easily eroded.

Current building design criteria for pole-house construction under the flood insurance program are outlined in the book *Elevated Residential Structures*. Regardless of insurance, pole-type construction with deep embedment of the poles is best in areas where waves and storm surge will erode foundation material. Materials used in pole construction include the following:

Piles. These are long, slender columns of wood, steel, or concrete driven into the earth to a sufficient depth to support the vertical load of the house and to withstand horizontal forces of flowing water, wind, and water-borne debris. Pile construction is especially suitable in areas where scouring (soil "washing out" from under the foundation of a house) is a problem.

Posts. Usually posts are made of wood; if of steel, they are called columns. Unlike piles, they are not driven into the ground, but rather are placed in a pre-dug hole at the bottom of which may be a concrete pad (fig. 6.3). Posts may be held in place by backfilling and tamping earth, or by pouring concrete into the

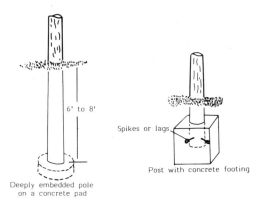

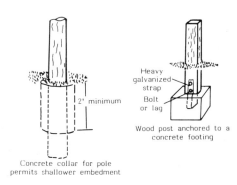

Fig. 6.3. Shallow and deep supports for poles and posts. Source: Southern Pine Association.

hole after the post is in place. Posts are more readily aligned than driven piles and are, therefore, better to use if poles must extend to the roof. In general, treated wood is the cheapest and most common material for both posts and piles.

Piers. These are vertical supports, thicker than piles or posts, usually made of reinforced concrete or reinforced masonry (concrete blocks or bricks). They are set on footings and extend to the underside of the floor frame.

Pole construction can be of two types. The poles can be cut off at the first-floor level to support the platform that serves as the dwelling floor. In this case, piles, posts, or piers can be used. Or they can be extended to the roof and rigidly tied into both the floor and the roof. In this way they become major framing members for the structure and provide better anchorage to the house as a whole (figs. 6.4 and 6.5). A combination of full- and floor-height poles is used in some cases, with the shorter poles restricted to supporting the floor inside the house (fig. 6.6).

Where the foundation material can be eroded by waves or winds, the poles should be deeply embedded and solidly anchored either by driving piles or by drilling deep holes for posts and putting in a concrete pad at the bottom of each hole. Where the embedment is shallow, a concrete collar around the poles improves anchorage (fig. 6.3). The choice depends on the soil conditions. Piles are more difficult than posts to align to match the house frame, as posts can be positioned in the holes before backfilling. Inadequate piling depths, improper piling-to-floor connections, and inadequate pile bracing all contribute to structural failure when storm waves liquify and erode sand support. Just as important as driving the

Fig. 6.4. Pole house, with poles extending to the roof. Extending poles to the roof, as shown in this photograph, instead of the usual method of cutting them off at the first floor, greatly strengthens a beach cottage. Photo by Orrin Pilkey, Jr.

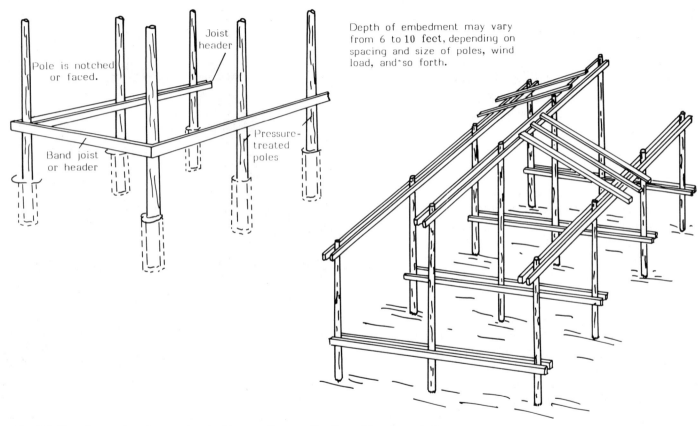

Joist header

Pole is notched or faced.

Band joist or header

Pressure-treated poles

Depth of embedment may vary from 6 to 10 feet, depending on spacing and size of poles, wind load, and so forth.

Fig. 6.5. Framing system for an elevated house. Source: Southern Pine Association.

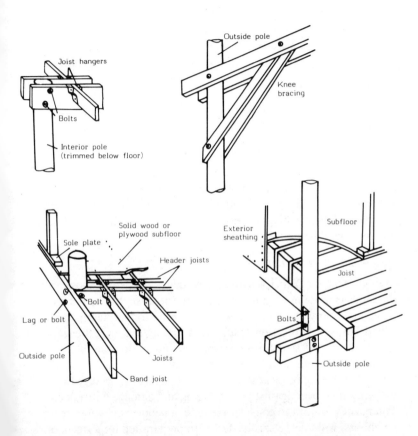

Fig. 6.6. Tying floors to poles. Source: Southern Pine Association.

piling deep enough to resist scouring and to support the loads they must carry is the need to fasten them securely to the structure they support above them. Unfortunately, many buildings on Florida's beaches according to local disaster officials are not so anchored. The connections must resist both horizontal loads from wind and wave during a storm and uplift from the same source.

When post holes are dug, rather than pilings driven, the posts should extend 4 to 8 feet into the ground to provide anchorage. The lower end of the post should rest on a concrete pad, spreading the load to the soil over a greater area to prevent settlement. Where the soil is sandy or is the type that the embedment can be less than, say, 6 feet, it is best to tie the post down to the footing with straps or other anchoring devices to prevent uplift. Driven piles should have a minimum penetration of 8 feet. However, most soils require greater embedment, as may code requirements for a specific situation. If the site is near the water, greater embedment is needed.

The floor and the roof should be securely connected to the poles with bolts or other fasteners. When the floor rests on poles that do not extend to the roof, attachment is even more critical. A system of metal straps is often used. Unfortunately, it sometimes happens that builders inadequately attach the girders, beams, and joists to the supporting poles by too few and undersized bolts. Hurricanes have proven this to be insufficient. During the next hurricane on the West Florida coast, many houses and condominiums will be destroyed because of inadequate attachment.

Local building codes may specify the size, quality, and spacing of the piles, ties, and bracing, as well as the methods of fastening the structure to them. Building codes often are minimal requirements, however, and building inspectors are usually amenable to allowing designs that are equally or more effective.

The space under an elevated house, whether pole-type or otherwise, must be kept free of obstructions in order to minimize the impact of waves and floating debris. If the space is enclosed, the enclosing walls should be designed so that they can break away or fall under flood loads, but also remain attached to the house or be heavy enough to sink. Thus, the walls cannot float away and add to the water-borne debris problem. Alternative ways of avoiding this problem are designing walls that can be swung up out of the path of the floodwaters, or building them with louvers that allow the water to pass through. The louvered wall is subject to damage from floating debris. The convenience of closing in the ground floor for a garage, storage area, or recreation room may be costly because it may violate insurance requirements and actually contribute to the loss of the house in a hurricane. The design of the enclosing breakaway walls should be checked against insurance requirements. See *Elevated Residential Structures* by the Federal Insurance Administration, Department of Housing and Urban Development, 451 Seventh Street, S.W., Washington, DC 20410, for more information.

An existing house: what to look for, what to improve

If instead of building a new house, you are selecting a house already built in an area subject to flooding and high winds, consider the following factors: (1) where the house is located; (2) how well the house is built; and (3) how the house can be improved.

Geographic location

Evaluate the site of an existing house using the same principles given earlier for the evaluation of a possible site to build a new house. House elevation, frequency of high water, escape route, and how well the lot drains should be emphasized.

You can modify the house after you have purchased it, but you cannot prevent hurricanes or other storms. The first step is to stop and consider; do the pleasure and benefits of this location balance the risk and disadvantage? If not, look elsewhere for a home; if so, then evaluate the house itself.

How well built is the house?

In general, the principles used to evaluate an existing house are the same as those used in building a new one. It should be remembered that many of the houses were built prior to the enactment of Flood Disaster Protection Insurance and may not meet the standards required of structures or improvements built since then.

Before you thoroughly inspect the building in which you are interested, look closely at the adjacent structures. If poorly built, they may float over against your building and damage it in a flood. You may even want to consider the type of people you will have as neighbors: will they "clear the decks" in preparation for a storm or will they leave items in the yard to become wind-borne missiles?

The house or condominium itself should be inspected for the following characteristics. The structure should be well anchored to the ground. If it is simply resting on blocks, rising water may cause it to float off its foundation and come to rest against your neighbor's house or out in the middle of the street. If well built and well braced internally, it may be possible to move the house back to its proper location, but chances are great that the house will be too damaged to be habitable.

If the building is on piles, posts, or poles, check to see if the floor beams are adequately bolted to them. If it rests on piers, crawl under the house if space permits to see if the floor beams are securely connected to the foundation. If the floor system rests unanchored on piers, do not buy the house.

It is difficult to discern whether a building built on a concrete slab is properly bolted to the slab because the inside and outside walls hide the bolts. If you can locate the builder, ask if such bolting was done. Better yet, if you can get assurances that construction of the house complied with the provisions of a building code serving the needs of that particular region, you can be reasonably sure that all parts of the house are well anchored: the foundation to the ground; the floor to the foundation; the walls to the floor; and the roof to the walls (figs. 6.7, 6.8, and 6.9). Be aware that many builders, carpenters, and building inspectors who are accustomed to traditional construction are apt to regard metal connectors, collar beams, and other such devices as newfangled and unnecessary. If consulted, they may assure you that a house is as solid as a rock, when in fact, it is far from it. Nevertheless, it

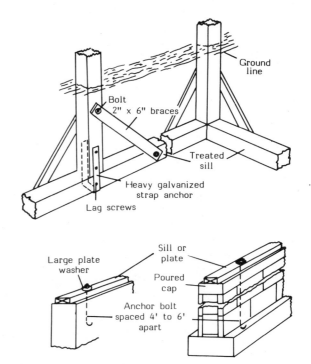

Fig. 6.7. Foundation anchorage. Top: anchored sill for shallow embedment. Bottom: anchoring sill or plate to foundation. Source of bottom drawing: *Houses Can Resist Hurricanes*, U.S. Forest Service Research Paper FPL 33.

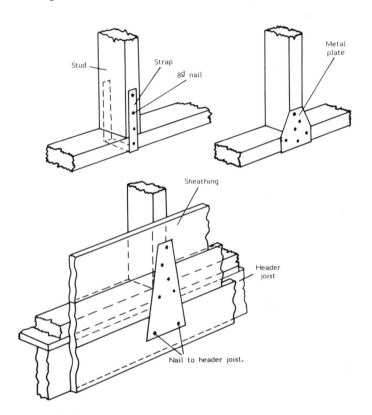

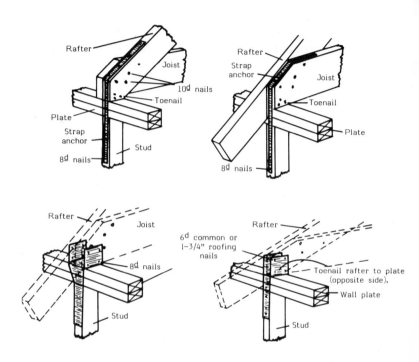

Fig. 6.8. Stud-to-floor, plate-to-floor framing methods. Source: *Houses Can Resist Hurricanes*, U.S. Forest Service Research Paper FPL 33.

Fig. 6.9. Roof-to-wall connectors. The top drawings show metal strap connectors: left, rafter to stud; right, joist to stud. The bottom left drawing shows a double-member metal plate connector—in this case with the joist to the right of the rafter. The bottom right drawing shows a single-member metal plate connector. Source: *Houses Can Resist Hurricanes*, U.S. Forest Service Research Paper FPL 33.

is wise to consult the builder or knowledgeable neighbors when possible.

The roof should be well anchored to the walls. This will prevent uplifting and separation from the walls. Visit the attic to see if such anchoring exists. Simple toe-nailing (nailing at an angle) is not adequate; metal fasteners are needed. Depending on the type of construction and the amount of insulation laid on the floor of the attic, these may or may not be easy to see. If roof trusses or braced rafters were used, it should be easy to see whether the various members, such as the diagonals, are well fastened together. Again, simple toe-nailing will not suffice. Some builders, unfortunately, nail parts of a roof truss just enough to hold it together to get it in place. A collar beam or gusset at the peak of the roof (fig. 6.10) provides some assurance of good construction. The Standard Building Code states that wood truss rafters shall be securely fastened to the exterior walls with approved hurricane anchors or clips.

Quality roofing material should be well anchored to the sheathing. A poor roof covering will be destroyed by hurricane-force winds, allowing rain to enter the house and damage ceilings, walls, and the contents of the house. Galvanized nails (2 per shingle) should be used to connect wood shingles and shakes to wood sheathing and should be long enough to penetrate through the sheathing (fig. 6.10). Threaded nails should be used for plywood sheathing. Sheathing is the covering (usually woodboards, plywood, or wallboards) placed over rafters, or exterior studding of a building, to provide a base for the application of roof or wall cladding. For roof slopes that rise 1 foot for every 3 feet or more of horizontal distance, exposure of the shingle should be about one-fourth of its length (4 inches for a 16-inch shingle). If shakes (thicker and longer than shingles) are used, less than one-third of their length should be exposed.

In hurricane areas, asphalt shingles should be exposed somewhat less than usual. A mastic or seal-tab type, or an interlocking shingle of heavy grade, should be used along with a roof underlay of asphalt-saturated felt and galvanized roofing nails or approved staples (6 for each 3-tab strip).

The fundamental rule to remember in framing is that all structural elements should be fastened together and anchored to the ground in such a manner as to resist all forces, regardless of which direction these forces may come from. This prevents overturning, floating off, racking, or disintegration.

The shape of the house is important. A hip roof, which slopes in 4 directions, is better able to resist high winds than a gable roof, which slopes in 2 directions. This was found to be true in Hurricane Camille in 1969 in Mississippi and, later, in Typhoon Tracy, which devastated Darwin, Australia, in December 1974. The reason is two-fold: the hip roof offers a smaller shape for the wind to blow against, and its structure is such that it is better braced in all directions.

Note also the horizontal cross section of the house (the shape of the house as viewed from above). The pressure exerted by a wind on a round or elliptical shape is about 60 percent of that exerted on the common square or rectangular shape; the pressure exerted

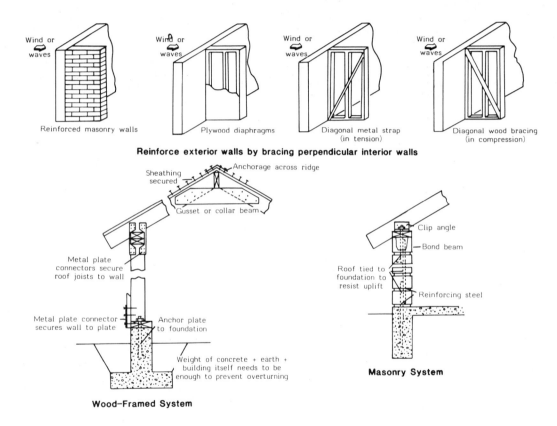

Reinforce exterior walls by bracing perpendicular interior walls

Reinforced masonry walls

Plywood diaphragms

Diagonal metal strap (in tension)

Diagonal wood bracing (in compression)

Wood-Framed System

Masonry System

Fig. 6.10. Where to strengthen a house. Modified from U.S. Civil Defense Preparedness Agency Publication TR-83.

on a hexagonal or octagonal cross section is about 80 percent of that exerted on a square or rectangular cross section (fig. 6.11).

The design of a house or building in a coastal area should minimize structural discontinuities and irregularities. It should be plain and simple and have a minimum of nooks and crannies and offsets on the exterior, because damage to a structure tends to concentrate at these points. Some of the newer beach cottages along the Florida shore are of a highly angular design with such nooks and crannies. Award-winning architecture will be a storm loser if the design has not incorporated the technology for maximizing structural integrity with respect to storm forces. When irregularities are absent, the house reacts to storm winds as a complete unit (fig. 6.11).

Brick, concrete-block, and masonry-wall houses should be adequately reinforced. This reinforcement is hidden from view. Building codes applicable to high-wind areas often specify the type of mortar, reinforcing, and anchoring to be used in construction. If you can get assurance that the house was built in compliance with a building code designed for such an area, consider buying it. At all costs, avoid unreinforced masonry houses.

A poured-concrete bond beam at the top of the wall just under the roof is one indication that the house is well built (fig. 6.12). Most bond beams are formed by putting in reinforcing and pouring concrete in U-shaped concrete blocks. From the outside, however, you cannot distinguish these U-shaped blocks from ordinary ones and therefore cannot be certain that a bond beam exists. The vertical reinforcing should penetrate the bond beam.

Some architects and builders use a stacked bond (1 block directly above another) rather than overlapped or staggered blocks because they believe it looks better. The stacked bond is definitely weaker than the latter. Unless you have proof that the walls are adequately reinforced to overcome this lack of strength, you should avoid this type of construction.

In past hurricanes the brick veneer of many houses has separated from the wood frame, even when the houses remained standing. Asbestos-type outer wall panels used on many houses in Darwin, Australia, were found to be brittle and broke up under the impact of wind-borne debris in Typhoon Tracy. Both types of construction should be avoided along the coast.

Ocean-facing glazing. Windows, glass doors, and glass panels should be minimal. Although large open glass areas facing the ocean provide an excellent sea view, such glazing may present several problems. The obvious hazard is disintegrating and inward-blowing glass during a storm. Glass projectiles are lethal. Less frequently recognized problems include the fact that glass may not provide as much structural strength as wood, metal, or other building materials; and ocean-facing glass is commonly damaged through sediment sand blasting, transported by normal coastal winds. The solution to this latter problem may be in reducing the amount of glass in the original design, or installing storm shutters which come in a variety of materials from steel to wood.

Consult a good architect or structural engineer for advice if you are in doubt about any aspects of a house. A few dollars spent for wise counsel may save you from later financial grief.

To summarize, a beach house should have: (1) roof tied to walls,

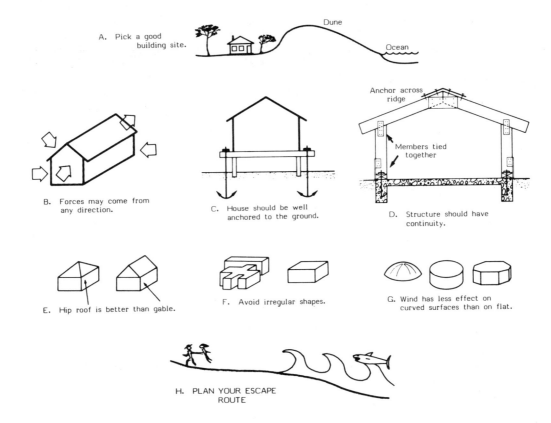

A. Pick a good building site.

B. Forces may come from any direction.

C. House should be well anchored to the ground.

Anchor across ridge

Members tied together

D. Structure should have continuity.

E. Hip roof is better than gable.

F. Avoid irregular shapes.

G. Wind has less effect on curved surfaces than on flat.

H. PLAN YOUR ESCAPE ROUTE

Fig. 6.11. Some rules in selecting or designing a house.

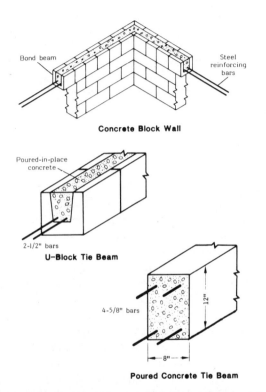

Concrete Block Wall

Poured-in-place concrete

2-1/2" bars
U-Block Tie Beam

4-5/8" bars
12"
8"
Poured Concrete Tie Beam

Bond beam

Steel reinforcing bars

Fig. 6.12. Reinforced tie beam (bond beam) for concrete block walls—to be used at each floor level and at roof level around the perimeter of the exterior walls.

walls tied to foundation, and foundation anchored to the earth (the connections are potentially the weakest link in the structural system); (2) a shape that resists storm forces; (3) floors high enough (sufficient elevation) to be above most storm waters (usually the 100-year flood level plus 3 to 8 feet); (4) piles or posts that are of sufficient depth or embedded in concrete to anchor the structure and to withstand erosion; and (5) piling that is well braced.

What can be done to improve an existing house?

If you presently own a house or are contemplating buying one in a hurricane-prone area, you will want to know how to improve occupant protection in the house. If so, you should obtain the excellent publication, *TR83 Wind Resistant Design Concepts for Residences*, by Delbart B. Ward, reference 64 (appendix C). Of particular interest are the sections on building a refuge shelter module within a residence. Also noteworthy are 2 supplements to this publication, *TR83A* and *TR83B* which deal with buildings larger than single-family residences in urban areas. These provide a means of checking whether the responsible authorities are doing their jobs to protect schools, office buildings, and apartments. Several pertinent references are listed in the bibliography (appendix C).

Suppose your house is resting on blocks but not fastened to them and, thus, is not adequately anchored to the ground. Can anything be done? One solution is to treat the house like a mobile home by screwing ground anchors into the ground to a depth of 4 feet or more and fastening them to the underside of the floor systems. See figures 6.13 and 6.14 for illustrations of how ground anchors can be used.

Calculations to determine the needed number of ground anchors will differ between a house and a mobile home, because each is affected differently by the forces of wind and water. Note that recent practice is to put these commercial steel-rod anchors in at an angle in order to better align them with the direction of the pull. If a vertical anchor is used, the top 18 inches or so should be encased in a concrete cylinder about 12 inches in diameter. This prevents the top of the anchor rod from bending or slicing through the wet soil from the horizontal component of the pull.

Diagonal struts, either timber or pipe, also may be used to anchor a house that rests on blocks. This is done by fastening the upper ends of the struts to the floor system, and the lower ends to individual concrete footings substantially below the surface of the ground. These struts must be able to take both uplift (tension) and compression and should be tied into the concrete footing with anchoring devices such as straps or spikes.

If the house has a porch with exposed columns or posts, it should be possible to install tiedown anchors on their tops and bottoms. Steel straps should suffice in most cases.

When accessible, roof rafters and trusses should be anchored to the wall system. Usually the roof trusses or braced rafters are sufficiently exposed to make it possible to strengthen joints (where 2 or more members meet) with collar beams or gussets, particularly at the peak of the roof (fig. 6.10).

A competent carpenter, architect, or structural engineer can review the house with you and help you decide what modifications are most practical and effective. Do not be misled by someone who is resistant to new ideas. One builder told a homeowner, "You don't want all those newfangled straps and anchoring devices. If you use them, the whole house will blow away, but if you build in the usual manner [with members lightly connected], you may lose only part of it."

In fact, the very purpose of the straps is to prevent any or all of the house from blowing away. The Standard Building Code says, "Lateral support securely anchored to all walls provides the best and only sound structural stability against horizontal thrusts, such as winds of exceptional velocity." And the cost of connecting all elements securely adds very little to the cost of the frame of the dwelling, usually under 10 percent, and a very much smaller percentage of the total cost of the house.

If the house has an overhanging eave and there are no openings on its underside, it may be feasible to cut openings and screen them. These openings keep the attic cooler (a plus in the summer) and help to equalize the pressure inside and outside the house during a storm with a low-pressure center.

Another way a house can be improved is to modify 1 room so that it can be used as an emergency refuge in case you are trapped in a major storm. (This is *not* an alternative to evacuation prior to a hurricane.) Examine the house and select the best room to stay in during a storm. A small, windowless room such as a bathroom, utility room, den, or storage space is usually stronger than a room with windows. A sturdy inner room, with more than one wall

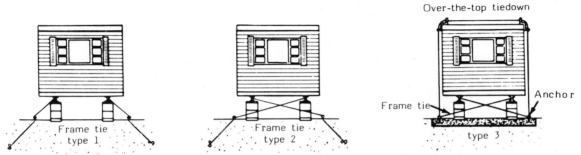

These sketches illustrate various methods for connecting frame ties to the mobile home frame. Type 2 system can resist greater horizontal forces than type 1. Type 3 system involves placement of mobile home on concrete slab. Anchors embedded in concrete slab are connected to ties.

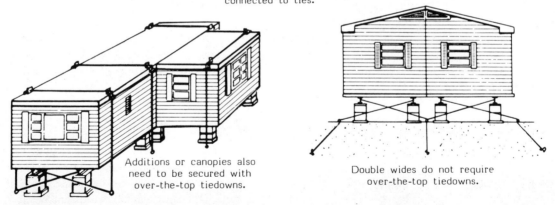

Fig. 6.13. Tiedowns for mobile homes. Source: U.S. Civil Defense Preparedness Agency Publication TR-75.

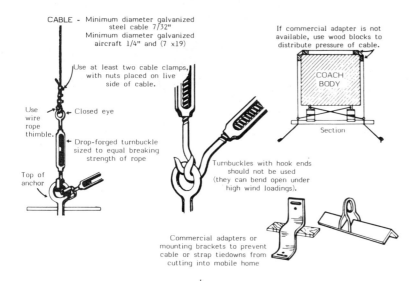

CABLE - Minimum diameter galvanized steel cable 7/32"
Minimum diameter galvanized aircraft 1/4" and (7 x19)

Use at least two cable clamps, with nuts placed on live side of cable.

If commercial adapter is not available, use wood blocks to distribute pressure of cable.

COACH BODY

Section

Use wire rope thimble.

Closed eye

Drop-forged turnbuckle sized to equal breaking strength of rope

Top of anchor

Turnbuckles with hook ends should not be used (they can bend open under high wind loadings).

Commercial adapters or mounting brackets to prevent cable or strap tiedowns from cutting into mobile home

Fig. 6.14. Hardware for mobile home tiedowns. Modified from U.S. Civil Defense Preparedness Agency Publication TR-75.

between it and the outside, is safest. The fewer doors, the better; an adjoining wall or baffle wall shielding the door adds to the protection.

Consider bracing or strengthening the interior walls. Such reinforcement may require removing the surface covering and installing plywood sheathing or strap bracing. Where wall studs are exposed, bracing straps offer a simple way to achieve needed reinforcement against the wind. These straps are commercially produced and are made of 16-gauge galvanized metal with prepunched holes for nailing. These should be secured to studs and wall plates as nail holes permit (fig. 6.10). Bear in mind that they are good only for tension.

If, after reading this, you agree that something should be done to your house, do it now. Do not put it off until the next hurricane hits you!

Mobile homes: limiting their mobility

Because of their light weight and flat sides, mobile homes are vulnerable to the high winds of hurricanes, tornadoes, and severe storms. Such winds can overturn unanchored homes or smash them into neighboring homes and property. Nearly 6 million Americans live in mobile homes today, and the number is growing. Twenty to 30 percent of single-family housing production in the United States consists of mobile homes. High winds damage or destroy nearly 5,000 of these homes every year, and the number will surely rise unless protective measures are taken. As one man whose mobile home was overturned in Hurricane Frederic (1979) so aptly put it, "People who live in flimsy houses shouldn't have hurricanes."

Several lessons can be learned from past experiences in storms. First, mobile homes should be located properly. After Hurricane Camille (1969), it was observed that where mobile home parks were surrounded by woods and where the units were close together, damage was minimized, caused mainly by falling trees. In unpro-

Table 6.1. Tiedown anchorage requirements

Wind velocity (mph)	10- and 12-ft.-wide mobile homes				12- and 14-ft.-wide mobile homes, 60 to 70 ft. long	
	30 to 50 ft. long		50 to 60 ft. long			
	No. of frame ties	No. of over-the-top ties	No. of frame ties	No. of over-the-top ties	No. of frame ties	No. of over-the-top ties
70	3	2	4	2	4	2
80	4	3	5	3	5	3
90	5	4	6	4	7	4
100	6	5	7	5	8	6
110	7	6	9	6	10	7

tected areas, however, many mobile homes were overturned and often destroyed from the force of the wind. The protection afforded by trees is greater than the possible damage from falling limbs. Two or more rows of trees are better than a single row, and trees 30 feet or more in height gives better protection than shorter ones. If possible, position the mobile home so that the narrow side faces the prevailing winds.

Locating a mobile home in a hilltop park will greatly increase its vulnerability to the wind. A lower site screened by trees is safer from the wind, but it should be above storm-surge flood levels. A location that is too low obviously increases the likelihood of flooding. There are fewer safe locations for mobile homes than for stilt houses.

A second lesson taught by past experience is that the mobile home must be tied down or anchored to the ground so that it will not overturn in high winds (figs. 6.13 and 6.14 and table 6.1).

Simple prudence dictates the use of tiedowns, and in Florida tiedowns are required. Many insurance companies, moreover, will not insure mobile homes unless they are adequately anchored with tiedowns. A mobile home may be tied down with cable or rope, or rigidly attached to the ground by connecting it to a simple wood-post foundation system. An alert mobile home park owner can provide permanent concrete anchors or piers to which hold-down ties may be fastened. In general, an entire tiedown system costs only a nominal amount.

A mobile home should be properly anchored with both ties to the frame and over-the-top straps; otherwise it may be damaged by sliding, overturning, or tossing. The most common cause of major damage is the tearing away of most or all of the roof. When this happens the walls are no longer adequately supported at the top and are more prone to collapse. Total destruction of a mobile home is more likely if the roof blows off, especially if the roof blows off first and then the home overturns. The necessity for anchoring cannot be overemphasized: there should be over-the-top tiedowns to resist overturning and frame ties to resist sliding off the piers. This applies to single mobile homes up to 14 feet in width. "Double-wides" do not require over-the-top ties but they do require frame ties. Although newer mobile homes are equipped with built-in straps to aid in tying down, the occupant may wish to add more if in a particularly vulnerable location. Many of the older mobile homes are not equipped with these built-in straps. *Protecting Mobile Homes from High Winds* treats the subject in more detail (see reference 70, appendix C). The booklet lists specific steps that one should take on receiving a hurricane warning and suggests a type of community shelter for a mobile home park. It also includes a map of the United States with lines that indicate areas subject to the strongest sustained winds. In a great hurricane, mobile homes will be destroyed no matter what you do to protect them.

High-rise buildings: the urban shore

A high-rise building on the beach is generally designed by an architect and a structural engineer who are presumably well qualified and aware of the requirements for building on the shoreline. Tenants of such a building, however, should not assume that it is therefore invulnerable. Many people living in apartment buildings of 2 or 3 stories were killed when the buildings were destroyed by Hurricane Camille in Mississippi in 1969. Storms have smashed 5-story buildings in Delaware. Larger high-rises have yet to be thoroughly tested by a major hurricane.

The first aspect of high-rise construction that a prospective apartment dweller or condo owner must consider is the type of piling used. High-rises near the beach should be built so that even if the foundation is severely undercut during a storm the building will remain standing. It is well known in construction circles that shortcuts are sometime taken by less scrupulous builders, and piling is not driven deeply enough. Just as important as driving the piling deep enough to resist scouring and to support the loads they must carry is the need to fasten piles securely to the structure they support above them. The connections must resist horizontal loads from wind and wave during a storm as well as uplift from the same sources. It is a joint responsibility of builders and building inspectors to make sure the job is done right. Hurricane Eloise (1975) exposed the foundation of a just-under-construction high-rise in Panama City Beach, revealing that 30 of the pilings had no concrete around them and were not attached to the building. Such problems probably exist everywhere that high-rises crowd the beach.

Despite the assurances that come with an engineered structure, life in a high-rise building holds definite drawbacks that prospective tenants should take into consideration. The negative conditions that must be evaluated stem from high wind, high water, and poor foundations.

Pressure from the wind is greater near the shore than it is inland, and it increases with height. If you are living inland in a 2-story house and move to the eleventh floor of a high-rise on the shore, you should expect 5 times more wind pressure than you are accustomed to. This can be a great and possibly devastating surprise.

The high wind pressure actually can cause unpleasant motion of the building. It is worthwhile to check with current residents of a high-rise to find out if it has undesirable motion characteristics; some have claimed that the swaying is great enough to cause motion sickness. More seriously, high winds can break windows and damage property, and of course they can hurt people. Tenants of severely damaged buildings will have to relocate until repairs are made.

Those who are interested in researching the subject further — even the knowledgeable engineer or architect who is engaged to design a structure near the shore — should obtain a copy of *Structural Failures: Modes, Causes, Responsibilities* (reference 62, appendix C). Of particular importance is the chapter entitled "Failure of Structures Due to Extreme Winds." This chapter analyzes wind damage to engineered high-rise buildings from the storm at Lubbock and Corpus Christi, Texas, in 1970.

Another occurrence that affects a multi-family, high-rise building more seriously than a low-occupancy structure is a power failure or blackout. Such an occurrence is more likely along the coast than inland because of the more severe weather conditions associated with coastal storms. A power failure can cause great distress. People can be caught between floors in an elevator. New York City had that experience once on a large scale. Think of the mental and physical distress after several hours of confinement, and compound this with the roaring winds of a hurricane whipping around the building, sounding like a freight train. In this age of electricity, it is easy to imagine many of the inconveniences that can be caused by a power failure in a multi-story building.

Fire is an extra hazard in a high-rise building. Even recently constructed buildings seem to have difficulties. The television pictures of a woman leaping from the window of a burning building in New Orleans rather than be incinerated in the blaze are a horrible reminder from recent history. The number of hotel fires of the last few years demonstrate the problems. Fire department equipment reaches only so high. And many areas along the coast are too sparsely populated to afford high-reaching equipment.

Fire and smoke travel along ventilation ducts, elevator shafts, corridors, and similar passages. The situation *can be* corrected and the building made safer, especially if it is new. Sprinkler systems should be operated by gravity water systems rather than by powered pumps (because of possible power failure). Gravity systems use water from tanks higher up in the building. Battery-operated emergency lights that come on only when the other lights fail, better fire walls and automatic sealing doors, pressurized stairwells, and emergency-operated elevators in pressurized shafts will all contribute to greater safety. Unfortunately, all of these improvements cost money, and that is why they are often omitted unless required by the building code.

There are 2 interesting reports on damage caused by Hurricane Eloise, which struck the Florida Panhandle the morning of September 23, 1975. One is by Herbert S. Saffir, a Florida consulting engineer; the other is by Bryon Spangler of the University of Florida. The forward movement of the hurricane was unusually fast, causing its duration in a specific area to be lessened, thus minimizing damage from both wind and tidal surge. The still-water height at Panama City was 16 feet above mean sea level, plus about a 3-foot topping wave. Wind gusts of 154 mph for a period of one-half hour were measured.

At least one-third of the older structures in the Panama City area collapsed. These were beach-front motels, restaurants, apartments, condominium complexes, and some permanent residences. The structures built on piling survived with minimal damage. In one case, part of a motel was on spread footings and part on piles. Just the part on spread footings was severely damaged. (A spread footing is a wide concrete slab resting directly on the ground rather than on piles.)

The anchorage systems, connection between concrete piles or concrete piers and the grade beams, under several high-rise buildings were inadequate to resist uplift loads, illustrating that code enforcement and proper inspection by a qualified professional are essential.

Many of the residences and some of the buildings were built on spread footings that failed because the sand they were resting on washed away with scour. Failure of the footings resulted in failure of the superstructure.

Some of the high-rise buildings suffered glass damage in both windows and sliding glass doors.

Apparently few, if any, of the residences and buildings were built to conform to either the South Florida or the Standard Building Code requirements. (The code was not legally applicable at the time.) If the requirements had been met, much of the damage could have been prevented at a minimum of cost.

Some surprising things were noticed. In almost every case where there was a swimming pool, considerable erosion occurred. Loss of sand beneath the walkways prior to the storm created a channel for the water to flow through and wash out more sand during the storm, which in turn increased both the velocity and quantity of the flow of water in the channel. This ate away the sand supporting adjacent structures, accelerating their failure.

Slabs on grade (on the ground) performed poorly. Often wave action washed out the sand underneath the slab. When this occurred there was no longer support for the structure above it, and failure resulted.

The storms revealed some shoddy construction. Some builders had placed wire mesh for a slab directly on the sand. Then the concrete was poured on top of it, leaving the mesh below and in the sand, where it served no structural purpose. To be effective, it should have been set on blocks or chairs, or pulled up into the slab during the pouring of the concrete.

In some cases cantilevered slabs or overhangs were reinforced for the usual downward gravity loads. Unfortunately, when waves dashed against the buildings they splashed upward, imposing an upward force against the slab for which it was not reinforced, causing it to crack and fail.

Modular unit construction: prefabricating the urban shore

The method of building a house, duplex, or large condominium structure by fabricating modular units in a shop and assembling them at the site is gaining in popularity for development on shoreline property. The larger of these structures are commonly 2 to 3 stories in height and may contain a large number of living units.

Modular construction makes good economic sense, and there is nothing inherently wrong in this approach to coastal construction. These methods have been used in the manufacturing of mobile homes for years, although final assembly on mobile homes is done in the shop rather than on the site. Doing as much of the work as possible in a shop can save considerable labor and cost. The workers are not affected by outside weather conditions. They often can be paid by piecework, enhancing their productivity. Shop work lends itself to labor-saving equipment such as pneumatic nailing guns and overhead cranes.

If the manufacturer desires it, shop fabrication can permit higher quality. Inspection and control of the whole process are much easier. For instance, there is less hesitation about rejecting a poor piece of lumber when you have a nearby supply of it than if you are building a single dwelling and have just so much lumber on the site.

On the other hand, because so much work is done out of sight of the buyer, there is the opportunity for the manufacturer to take shortcuts if he is so inclined. It is possible that some modular dwelling units have their wiring, plumbing, ventilation, and heating and air conditioning installed at the factory by unqualified personnel, and it is possible the resulting inferior work is either not inspected or inspected by an unconscientious or inept individual.

Therefore, it is important to consider the following: Were wiring, plumbing, heating and air conditioning, and ventilation installed at the factory or at the building site? Were the installers licensed and certified? Was the work inspected at both the factory and on the construction site? Most important, is the modular dwelling unit built to provide safety in the event of fire? For example, just a few of the many safety features that should be included are 2 or more exits, stairs remote from each other, masonry fire walls between units, noncombustible wall sheeting, and compartmentalized units so that if fire does occur it will be confined to that 1 unit.

It is vital that if 1 unit is placed on top of another, they be adequately fasten together to resist high winds and that they not depend solely on the weight of the upper unit to hold it in place.

In general, it is very desirable to check the reputation and integrity of the manufacturer just as you would when hiring a contractor to build your individual house on site. The acquisition of a modular unit should be approached with the same caution as for other structures.

If you are contemplating purchasing one of these modularized dwelling units, you may be well advised to take the following steps:

1. Check the reputation and integrity of the developer and manufacturer.
2. Check to see if the developer has a state contractor's license.

3. Check the state law on who is required to approve and certify the building.
4. Check what building codes are enforced.
5. Check to see if the state fire marshal's office has indicated that the dwelling units comply with all applicable codes. Also check to see if this office makes periodic inspections.
6. Check to see that smoke alarms have been installed, if windows are the type that can be opened, if the bathroom has an exhaust fan, and if the kitchen has a vent through the roof.

As with all other types of structures, also consider site safety and escape routes for the location of modular units.

An unending game: only the players change

Hurricane or calm, receding shore or growing shore, storm-surge flood or sunny sky, migrating dune or maritime forest, win or lose, the gamble of coastal development will continue. If you choose your site with natural safety in view, follow structural engineering design in construction, and take a generally prudent approach to living at the shore (fig. 6.11), then you become the gambler who knows when to hold them, when to fold them, and when to walk away.

Our goal is to provide guidance to today's and tomorrow's players. This book is not the last nor by any means the complete guide to coastal living, but it should provide a beginning. In the appendixes that follow chapter 7 are additional resources that we hope every reader will pursue.

7. The coast, land use, and the law

The previous chapters demonstrate that the coastal zone is a dynamic area where land, wind, wave, and organisms intereact. The resulting rapid changes are especially apparent on barrier islands. We cannot build and live in this zone without some level of interference, or without risking the negative impacts brought about by natural changes.

Coastal dynamics preclude shoreline and island development patterned after traditional inland styles. A 1-story, ranch-style house at the back of the beach will block wind transport of sand, interfere with overwash, and ultimately behave as a seawall before being destroyed in its turn by storm waves and flooding. This traditional design in this dynamic zone would have a much shorter life expectancy than the same house in an inland location. The services for this house and many like it (for example, electric lines, gas mains, water lines), the sewage generated, and the roads, bridges, and service structures required for such development will exceed the carrying capacity of a barrier island much quicker than for a similar inland community. The resulting damage to the environment through pollution, loss of habitat, stabilization structures, and the like removes the amenities that most shore dwellers originally came to enjoy. Not only aesthetic value is lost, but the risk from coastal hazards is increased

Wise land-use planning, environmental maintenance, and conservation of the coastal zone are necessary to protect the environment. But just as significant, they are necessary to protect ourselves. The ecosystem is as important to the human population as it is to a population of pelicans or a stand of sea oats. Curiously, laws are passed to protect the latter with the goal of protecting the former—sometimes from ourselves.

Population growth, affluence, and migration to the Sun Belt will necessitate increased regulation of the coastal zone. Florida's coastal population is expected to approach 10 million by the year 2000! By analogy, as the traffic increases, more traffic laws and regulations are required to avoid the certainty of traffic jams.

The best philosophy on shoreline development is that land use should be in harmony with the natural environments and processes that constitute the system. Of course, various segments of society view the coastal zone differently. The extreme views range from untouched preservationism to unplanned, uncontrolled urbanization. Increasingly, decisions on land use are made by governments under the pressure of various special interest groups. Existing legislation is often that of compromise, satisfying the various federal, state, and local levels of the political infrastructure. We can expect that regulations will continue to be established and modified with the intention of insuring reasonable, multiple land use of the coastal zone, while attempting to protect both inhabitants and the natural environment. Developers have had this expectation in the past, and in some cases it has spurred unwise development.

That is, buildings have been constructed before tighter restrictions could go into effect. Current and prospective owners of coastal property, especially on barrier islands, should be aware of their responsibilities under current law and expect additional regulation with respect to development and land use.

A partial list of relevant current land-use programs and regulations applicable to the Florida coast follows. The explanations provided are general and introductory in nature; appendix B lists the agencies that will supply more specific and detailed information. The regulations listed here range from federal laws that protect the interests of the larger society to state and local laws and ordinances that serve the interests of Florida citizens and the local community. A review of these regulations before investing in or undertaking property development anywhere on the coast will be in your best interest. We recommend that you contact your local county or municipal planning, zoning, or building departments to determine state and federal permit requirements.

Coastal Barrier Resources Act of 1982

Recognizing the serious hazards, costs, and problems with federally subsidized development of barrier islands, the U.S. Congress passed the Coastal Barrier Resources Act (Public Law 97-348) in October 1982. The purpose of this federal law is to minimize loss of human life and property, wasteful expenditure of federal taxes, and damage to fish, wildlife, and other natural resources from incompatible development along the Atlantic and Gulf coasts. The act covers 190 designated areas, covering 700 miles of undeveloped barrier beaches in the United States.

Specifically, the act prohibits the expenditures of federal funds, including loans and grants, for the construction of infrastructures that encourage barrier island development; these infrastructures include roads, bridges, water supply systems, waste water treatment systems, and erosion control projects. Any new structure built on these designated barrier islands after October 1, 1983, is not eligible for federal flood insurance. Certain activities and expenditures under the act are permissible. The act does not prohibit private development on the designated barrier islands, but it passes the risks and costs of development from taxpayers to owners. All applicable federal, state, and local permits still must be obtained before any development begins in the designated areas.

The Coastal Barrier Resources Act affects all or part of 33 of Florida's barrier islands covering 110 miles of ocean beaches. For exact boundaries of the designated areas, contact local city or county planning departments or the Florida Coastal Zone Management Program Office (appendix B).

National Flood Insurance Program (NFIP)

Florida's barrier beaches are prone to flood damage from hurricanes and tropical storms. The probability of a hurricane striking Florida's coastline is very high, up to 1 hurricane every 1.5 years. During hurricanes, storm surge and wave heights reach 12 to 20 feet above normal, and winds of 100 to 150 miles per hour are not uncommon. Between 1900 and 1980, hurricanes inflicted approximately $1.5 billion worth of damage in the Florida coastal zone. However, this figure is misleading because it is not in terms of today's inflated dollar. Given the present heavily developed Florida coasts, a *single* moderate hurricane could match or exceed this level of destruction!

The National Flood Insurance Act of 1968 (P.L. 90-448) as amended by the Flood Disaster Protection Act of 1973 (P.L. 92-234) was passed to encourage prudent land-use planning and to minimize property damage in flood-prone areas like barrier beaches. Local communities must adopt ordinances to reduce future flood risks in order to qualify for the National Flood Insurance Program. The NFIP provides an opportunity for property owners to purchase flood insurance that generally is not available from private insurance companies.

The initiative for qualifying for the program rests with the community, which must contact the Federal Emergency Management Agency (FEMA). FEMA will provide the community with a Flood Hazard Boundary Map (FHBM). Any community may join the National Flood Insurance Program provided that it requires development permits for all proposed construction and other development within the flood zone and ensures that construction materials and techniques are used to minimize potential flood damage. At this point the community is in the "emergency phase" of the NFIP. The federal government makes a limited amount of flood insurance coverage available, charging subsidized premium rates for all existing structures and/or their contents,

regardless of the flood risk.

FEMA may provide a more detailed Flood Insurance Rate Map (FIRM) indicating flood elevations and flood-hazard zones, including velocity zones (V-zones) for coastal areas where wave action is an additional hazard during flooding. The FIRM identifies Base Flood Elevations (BFEs), establishes special flood-hazard zones, and provides a basis for floodplain management and the establishing of insurance rates.

To enter the regular program phase of the NFIP, the community must adopt and enforce floodplain management ordinances that at least meet the minimum requirements for flood-hazard reduction as set by FEMA. The advantage of entering the regular program is that increased insurance coverage is made available, and new development will be more hazard-resistant. All new structures will be rated on an actual risk (actuarial) basis, which may mean higher insurance rates in coastal high-hazard areas but generally results in a savings for development within numbered A-zones (areas flooded in a 100-year coastal flood, but less subject to turbulent wave action).

FEMA maps commonly use the 100-year flood as the BFE to establish regulatory requirements. Persons unfamiliar with hydrologic data sometimes mistakenly take the 100-year flood to mean a flood that occurs once every 100 years. In fact, a flood of this magnitude could occur in successive years, or twice in one year, and so on. The flooding in Jackson, Mississippi, that has occurred over the last few years illustrates this point. If we think of a 100-year flood as a level of flooding having a 1 percent probability of occurring in any given year, then during the life of a house within this zone that has a 30-year mortgage, there is a 30 percent probability that the property will be flooded. The chance of losing your property becomes 1 in 4, rather than 1 in 100. Having flood insurance makes good sense.

In V-zones, new structures will be evaluated on their potential to withstand the impact of wave action, a risk factor over and above the flood elevation. Elevation requirements are adjusted, usually 3 to 6 feet above still-water flood levels, for structures in V-zones to minimize wave damage, and the insurance rates also are higher. When your insurance agent submits an application for a building within a V-zone, an elevation certificate that verifies the postconstruction elevation of the first floor of the building must accompany the application.

The insurance rate structure provides incentives of lower rates if buildings are elevated above the minimum federal requirements. General eligibility requirements vary among pole houses, mobile homes, and condominiums. Flood insurance coverage is provided for structural damage as well as contents. Table 7.1 presents Florida's coastal counties that are participating in the National Flood Insurance Program as of August 1981. Almost all coastal communities with barrier beaches are now covered under the regular program. To determine if your community is in the NFIP and for additional information on the insurance, contact your local property agent or call the NFIP's servicing contractor (phone: (800) 638-6620), or the NFIP State Assistance Office at (904) 488-9210. For more information, request a copy of "Questions and Answers on the National Flood Insurance Program" from FEMA (see appendix B under Insurance).

Before buying or building a structure on a barrier beach, an individual should ask certain basic questions:

1. Is the community I'm locating in covered by the emergency or regular phase of the National Flood Insurance Program?
2. Is my building site located in the designated areas of the Coastal Barrier Resources Act, where no federal flood insurance for new structures will be available after October 1, 1983.
3. Is my building site above the 100-year flood level? Is the site located in a V-zone? V-zones are high-hazard areas and pose serious problems.
4. What are the minimum elevation and structural requirements for my building?
5. What are the limits of coverage?

Make sure your community is enforcing the ordinance requiring minimum construction standards and elevations. After Hurricane Frederic (1979) a number of homeowners from Santa Rosa County, whose houses were flooded, put in claims for federal flood insurance. It developed that on direct order from the county commissioners the elevation requirements for insurance were not being enforced by the county. One woman who had paid $158 per year for her insurance discovered she should have been paying over $13,000 a year because her house was 5 feet below the 100-year flood level. Prior to construction, her house plans had been approved by the county and no mention was made of the elevation

Table 7.1. Flood insurance policies in coastal counties of Florida (as of August 31, 1981)

County	Regular (R) or emergency (E) program	Policies in effect	Dollar value of policies	County	Regular (R) or emergency (E) program	Policies in effect	Dollar value of policies
Counties with barrier islands and beaches				*Counties without barrier islands and beaches*			
Bay	R	4,005	222,805,400	Citrus	E	1,358	47,419,600
Brevard	R	18,294	1,419,068,800	Dixie	E	176	4,284,000
Broward	R	111,354	7,075,675,400	Hernando	E	625	20,663,600
Charlotte	R	8,865	423,679,600	Jefferson	E	12	297,800
Collier	R	13,800	785,170,200	Levy	E	231	7,211,600
Dade	R	95,424	5,130,950,800	Pasco	E	11,576	408,839,800
Duval	R	4,373	319,556,800	Taylor	E	111	3,010,300
Escambia	R	4,495	309,675,500	Subtotal		14,089	$ 491,726,700
Flagler	E	655	23,842,600				
Franklin	E	1,000	36,737,400	*Noncoastal counties*			
Gulf	E	389	11,117,000	Subtotal		23,171	$ 1,288,404,000
Hillsborough	R	13,447	788,578,800				
Indian River	R	4,801	43,623,400	Florida total		531,091	$ 30,563,259,400
Lee	R/E	33,608	1,346,771,400				
Manatee	R	10,549	577,315,000	18 Atlantic and Gulf states			
Martin	R	5,916	285,064,200	coastal counties total		1,164,798	$ 64,668,610,400
Monroe	R	17,034	789,522,200				
Nassau	R/E	753	37,482,800	All other states total		745,620	$ 33,303,788,600
Okaloosa	R/E	2,823	228,510,500				
Palm Beach	R	47,334	3,527,099,000	United States total		1,918,318	$ 97,972,399,000
Pinellas	R	57,019	3,132,368,200				
Santa Rosa	R	1,611	117,517,200				
Sarasota	R	16,467	889,767,600				
St. Johns	R	3,306	220,283,500				
St. Lucie	R/E	6,134	233,698,600				
Volusia	R	8,867	622,244,000				
Wakulla	R	352	11,446,400				
Walton	R	1,156	73,556,400				
Subtotal		493,831	$ 28,683,128,700				

Compiled by Dinesh C. Sharma.
Source: Federal Emergency Management Agency, Flood Insurance Administration, Washington, D.C. Personal communication, October 10, 1981.

problem. Before payment of her $17,000 claim, the National Flood Insurance Program subtracted her correct $13,000 premium. Later all parties agreed on a lower, but still substantial figure for flood insurance premiums. More than 20 people in the National Flood Insurance Program in the local community were forced to continue paying exorbitant insurance premiums for buildings built below the required elevation because the banks that held their mortgages insisted on it. All of this cost and confusion because county officials said nothing about flood elevations when issuing permits. The then-incumbent county commissioners fared very poorly in the next election!

Most lending institutions and community planning, zoning, and building departments will be aware of the flood insurance regulations and can provide assistance. It would be wise to confirm such information with appropriate insurance representatives. Any authorized insurance agent can write and submit a National Flood Insurance Program policy application. All insurance companies charge the same rates for national flood insurance policies.

The National Flood Insurance Program states its goal as "to . . . encourage state and local governments to make appropriate land use adjustments and to constrict the development of land which is exposed to flood damage and minimize damage caused by flood losses" and "to . . . guide the development of proposed future construction, where practical, away from locations which are threatened by flood hazard." To date, development in the flood-hazard areas continues at a rapid rate.

Revision of minimum flood elevations in the V-zones of coastal counties takes into account the additional hazard of storm waves atop still-water flood levels. Existing FEMA regulations stipulate protection of "dunes and vegetation" in the V-zones, but implementation of this requirement by the local communities has not always been strong. The existing requirements of the NFIP do not address other hazards of "migrating" shorelines, for example, shoreline erosion or shifting of inlets. Thus, buildings may meet the minimum FEMA elevation requirements but at the same time can be located near highly exposed and eroding shorelines. In addition to recognizing the flood hazard, the need exists to incorporate location and structural codes that reflect migrating shorelines, hurricane winds, wave uplift, horizontal pressures, and scouring to minimize the loss of structures as well as the dollars that have supported the insurance program. This is not to say that state and local codes and ordinances have overlooked the latter.

In the past the National Flood Insurance Program has been subsidized and has grown to become a large federal liability. As of August 31, 1981, more than 1.918 *million* flood insurance policies valued at $97.972 *billion* had been sold nationwide. Coastal counties had 1.165 million of these policies valued at $64.667 billion. Florida had more policies and coverage than any other state— 531,091 policies valued at $30.563 billion along the coastal counties with beaches (table 7.1). During 1978–79 the average premium for federal flood insurance policies located in velocity zones was $131 a year. Because of Hurricane Frederic, the average expense and loss per policy was $422, making it a costly subsidy for the nation's taxpayers. Such losses have encouraged the addition of requirements on wave heights to flood elevations and a major revision of the insurance rating system. As a result, insurance rates have been raised significantly.

Recognition of natural hazards and tax subsidy problems provided part of the rationale for Congress to pass the Coastal Barrier Resources Act in 1982. There is an urgent national need to address the problems of developed or developing barrier beaches that were not covered in the act in order to minimize hazards to human lives and loss of property in these areas. Incentive programs to encourage sound land-use planning, limit density of development, improve hurricane evacuation, and allow relocation of damaged structures after hurricanes need to be developed before a disaster hits the coast.

Tarpon Springs on the West Florida coast holds the distinction of having been the first community removed from the NFIP for not abiding by its agreements. Fortunately, the removal was only temporary and the community was reinstated to the program after improving its ordinance language and application forms, stiffening variance procedures, and upgrading the quality of the staff responsible for administering the program.

Clearly, FEMA is serious about enforcement. If a community is removed from the program, the result is that property owners cannot renew their flood insurance when it expires or buy new policies. Several other communities in Florida are probably in violation of flood insurance requirements and will face loss of insurability if they do not begin to enforce the program's requirements. One of the most common violations that local officials have ignored is the enclosing of ground-level portions of cottages and using them as parts of residences.

There are 2 ways the property owner is likely to get caught if he or she is in violation of the construction requirements. First, FEMA sends out inspectors periodically to see if communities are in compliance. Second, if you file a claim after storm flooding and damage, your property will be inspected. If your structure was in violation of construction requirements, you will be required to pay additional back premiums that could equal and even exceed the amount of the insurance claim. This has happened in Galveston Island, Texas.

Two points are clear. First, the property owner cannot rely solely on the developer, building inspector, or county commissioners to enforce the community ordinances required to qualify for and stay in the NFIP. Second, given the likelihood that developers will be long gone when the question of compliance arises, town and county officials are likely to be held responsible for the inaction in local enforcement and become the defendants in legal actions; that is, elected officials may lose more than the next election for not doing their mandated jobs. In California a group of homeowners who lost their houses in a landslide are suing local officials, claiming they were not warned of the hazard. A homeowner who loses his flood insurance coverage because the developer or community official was irresponsible is likely to take similar action.

Hurricane evacuation

The Disaster Relief Act of 1974 authorized FEMA to establish disaster preparedness plans in cooperation with local communities and states. Hurricane evacuation is a critical problem on barrier islands and coastal floodplains. Due to heavy concentrations of population in areas of low topography, narrow roads, and vulnerable bridges and causeways, plus limited hurricane warning capability (possibly 12 hours or less), it would be impossible to evacuate all of the people prior to hurricanes in many parts of Florida.

Several coastal communities in Florida have formulated detailed hurricane evacuation plans. You should check for hurricane evacuation plans with the county Civil Defense or Disaster Preparedness officer and find out if any potential evacuation problems will exist during a hurricane. They can provide information on the location of hurricane evacuation shelters. These same agencies are responsible for providing emergency and relocation assistance after hurricanes. The Civil Defense office also can provide information on expected losses from hurricanes.

The Florida Coastal Management Program (FCMP)

The Federal Coastal Zone Management Act of 1972 (CZMA) set in motion an effort by most coastal states to manage their shorelines and thereby conserve a vital national resource. Key requirements of the CZMA are coastal land-use planning based on land classification and identification and protection of critical areas. The intentions are to insure good land use and resource development, conserve resources, and protect the quality of life for citizens of the coastal zone.

While some states passed specific acts to set up state offices of coastal zone management, Florida established a program based on existing state laws. More than 20 statutes serve as the authorities for the Florida Coastal Management Program under the Florida Coastal Management Act of 1978 (Chapter 380, Florida Statutes), which was approved by the federal office of Coastal Zone Management (now the Office of Ocean and Coastal Resource Management) on September 24, 1981. The Department of Environmental Regulation is the designated coastal zone management agency, but it works closely with the Departments of Natural Resources and Community Affairs in implementing the program. The Interagency Management Committee (IMC), consisting of heads of state agencies involved in resource management, was established to solve complex coastal problems through joint efforts between agencies. In addition, the governor's Environment Land Management Study Committee (ELMS) and a legislative Growth Management Committee are reviewing Florida's environmental laws and may suggest legislative or administrative changes that could affect the Florida Coastal Management Program. For specific information, contact the Office of Coastal Zone Management (see appendix B: Coastal Zone Management).

Various aspects of the program are included under the following state programs and acts.

Hazard mitigation

Under the authority of the federal Disaster Relief Act of 1974

(P.L. 93–288) and Florida's Disaster Preparedness Act (Chapter 252, Florida Statutes), as well as other codes, the Bureau of Emergency Management is charged with responsibility for peacetime emergency planning. The purpose of emergency management is to improve public safety by protecting life and property in the event of natural or man-caused hazards. In the coastal zone these hazards include storms, hurricanes, flooding, overwash, shoreline erosion, erosion by shifting streams or channels (avulsion), including inlet migration, dune migration, pollution hazards, and so on (that is, the same hazards on which this book focuses). Hazard mitigation simply means reducing the likelihood of damage from such hazards through actions taken *before* the hazardous process occurs.

According to the state's Comprehensive Emergency Management Plan these are the long-term goals of the state's hazard mitigation effort:

Protection of life and property through the reduction and avoidance of unnecessary and uneconomical uses of hazardous areas.

Preservation and enhancement of beneficial uses of hazard-prone areas.

Protection of natural systems that serve a hazard moderating or mitigation function.

Attention is focused in particular on predictable, recurring hazards, like those noted above, and on seeking nonstructural solutions to hazard mitigation.

The Division of Public Safety Planning and Assistance, Department of Community Affairs, acts as a coordinating agency for developing policy, disseminating information on hazard mitigation, and making recommendations to other units of government. The agency also is responsible for site-specific hazard mitigation studies.

Community officials, planners, and individual property owners in the coastal zone should make use of the services of the Bureau of Emergency Management within the Division of Public Safety and Planning Assistance when evaluating site safety, seeking ways to reduce hazard impact, or planning strategies to meet hazard crises (for example, hurricane warning, evacuation, poststorm recovery).

Florida's Save Our Coast Program

The Coastal Barrier Resources Act of 1982 did not provide any funds for the acquisition of undeveloped barrier islands and beaches for public recreation, habitat protection, or hazard mitigation purposes. In 1981 Florida's governor and cabinet recognized the serious problems associated with the development of barrier islands and beaches such as loss of public access to beaches for recreation, economic losses due to severe erosion and flood damage, logistics for hurricane evacuation, disaster relief assistance costs, and subsidies for infrastructures. As a result, they started programs intended to protect and manage the barrier beaches. Under the $200 million Save Our Coast Program for acquiring undeveloped barrier island and beach properties, the state has a systematic process for the nomination and selection of parcels for purchase. This new program is in addition to the Conservation and Recreational Land (CARL) Program under which Florida acquires land for public use and recreation.

The governor of Florida signed an executive order to limit the expenditures of state funds for the construction of public infrastructures such as water and sewer systems, roads, bridges, and similar structures in certain hazardous coastal areas. If you plan to acquire or sell properties on a barrier island, it is advisable that you contact the governor's Office of Planning and Budgeting to determine if your property is located in one of the units where the executive order is applicable, or whether the state is interested in acquiring the land under the Save Our Coast Program or the CARL program.

Development of regional impacts

In 1972 the Florida Legislature enacted the Florida Environmental Land and Water Management Act of 1972 (Chapter 380, Florida Statutes) to address the problems of large-scale developments in the state. Under this law "any development because of its character, magnitude or location that would have substantial effect on the health, safety and welfare of citizens of more than one county" is considered a Development of Regional Impact (DRI). Types of projects include, but are not limited to, residential projects, tourist attraction and recreational facilities, shopping centers,

office buildings, parks, industrial parks, airports, port facilities, schools, and similar developments.

The procedures and rules pertaining to the determination and review of DRIs are contained in Florida Administrative Code, Rule 9B-16 and Rule 27F-1 Part II. Florida Administrative Code 27F-2 identifies those developments that are specifically presumed to be DRIs. The review and permit process is regulated by the Bureau of Land and Water Management of the Florida Department of Community Affairs (DCA). There are 11 Regional Planning Councils (RPCs) that actually conduct the DRI assessment and review (see appendix B for addresses). We recommend that you contact primarily the Bureau of Land and Water Management and also the appropriate regional planning council to determine if your project is a Development of Regional Impact. Projects located on barrier islands and beaches, around state aquatic preserves, or in environmentally designated areas are given closer scrutiny by the DCA and the RPCs.

Water pollution control and water supply

The Florida Air and Water Pollution Control Act of 1967 and subsequent amendments (Chapter 403, Florida Statutes) govern the discharge and regulation of domestic, municipal, and industrial water pollution. The state law incorporates the requirements of the Federal Water Pollution Control Act Amendments of 1972 and 1977. The Florida Department of Environmental Regulation (DER) has enforcement power over all natural or artificial bodies of water. The state has adopted a comprehensive set of water-quality standards. All waters in the state have been classified into 1 of the 7 classifications for the beneficial use of humans as well as propagation of fish, shellfish, and wildlife.

Water resources are being threatened with pollution, causing economic losses to both local communities and the state. In Florida one-third of all commercial shellfish harvesting areas have been permanently closed due to pollution! Protection of water quality is vital for human as well as other uses. State laws provide additional standards and protection of water quality if the waste disposal affects potable water supply (Class I), shellfish propagation and harvesting waters (Class II), aquatic preserves, or "Outstanding Florida Waters." You need to contact the Department of Environmental Regulation to find out if your property borders on environmentally sensitive waters.

If you plan to locate on a barrier island, check with the local county or city planning and building departments to see if adequate public drinking water supply and sewer hookups are available. In case you plan to build your own sewage disposal facility with a capacity of less than 2,000 gallons per day, your county health department must be contacted for a permit. Minimum standards for septic systems must be met. The first important standard is that the depth to the seasonally high groundwater table must be at least 3.0 feet below the bottom of the drainfield or about 5.0 feet below the ground surface in the wet season. The second important standard is that the drainfield must be set back a minimum of 50 feet from any surface water body. These mini-mum standards are intended to protect the public health and water quality. For any on-site sewage system or package treatment with a capacity greater than 2,000 gallons per day, contact the Florida DER for rules, regulations and guidelines.

The withdrawal or diversion of drinking water is governed by the Florida Water Resources Act of 1972 (Chapter 373, Florida Statutes). The responsibility for the enforcement of this law rests with the Florida Department of Environmental Regulation, but it is implemented through five water management districts (appendix B). If you obtain your water supply from a public facility or plan to put in a small domestic well, you need no permit from any water management district. However, if you plan to drain the land, divert the water, or put in a large water supply system, contact the appropriate water management district for rules, regulations, and guidelines.

On-site individual sewage disposal facilities

In many instances, development on barrier islands requires individual, on-site sewage disposal and treatment systems when a public sewer system is not available. The installation of on-site septic systems in Florida is regulated by the Department of Health and Rehabilitative Services (DHRS) pursuant to Section 381.272 of Florida Statutes. The local county health departments are responsible for the direct regulation and permitting of these facilities under Rule 10D-6 of Florida Administrative Codes. Rule 10D-6 has specific standards for the capacity of septic tanks, soil types, depth to seasonally high water table, minimum distance from drinking water wells or public water supply, and minimum distance from surface water bodies.

If you plan to install an individual, on-site sewage facility with an estimated daily flow of less than 2,000 gallons for any 1 establishment or structure, you must contact and obtain a permit from your county health department. Subdivisions of 50 or fewer lots, each having a minimum of at least 0.5 acre and a minimum dimension of 100 feet, may be developed with private wells and individual sewage disposal systems, provided satisfactory groundwater can be obtained, and all distance, setback, soil condition, water table elevations, and other requirements of Rule 10D–6 can be met. Residential subdivisions using public water supply systems may be developed with individual sewage disposal facilities for a maximum of 4 lots per acre, provided all other conditions are met.

The installation of septic systems on the highly permeable sandy soils of Florida barrier islands causes pollution of groundwaters as well as surface and estuarine waters. High groundwater tables during wet seasons make the septic systems a health hazard and vulnerable to failures. It is imperative that proliferation of on-site septic systems in highly permeable soils, close to the surface water bodies, and under seasonally high groundwater table conditions, be discouraged to protect public health and water quality.

Dredging and filling

Saltwater and freshwater wetlands are considered extremely valuable natural resources in Florida. Florida legislative goals and policies reflect this concern under Chapter 253 and Chapter 403, Florida Statutes. These policies state in part,

> to prohibit the authorization of the dredging and filling of submerged lands, if such authorization would result in the destruction of resources or interfere with public uses to such an extent as to be contrary to the public interest [and] to prevent and abate pollution and to conserve the waters of the state for the propagation of wildlife, fish and other aquatic life, and for domestic agricultural, industrial, recreational and other beneficial uses.

To minimize permit problems, delays, and frustrations, we suggest that you do not buy properties located in wetlands of barrier islands and coastal areas.

Barrier islands are characterized by the presence of freshwater and saltwater wetlands. If your plan requires any dredging and/or filling of wetlands or navigable waters, you need to obtain permits from the appropriate state and federal agencies. The dredging and/or filling activity may be associated with the construction of a homesite, access road, boat dock, or erosion control structure. Unauthorized dredging and filling is prohibited and punishable under state and federal laws.

The dredging and filling in Florida waters is governed by Florida Air and Water Pollution Control Act (Chapter 403, Florida Statutes), Florida State Land Trust Fund (Chapter 253, Florida Statutes) and Beach and Shore Preservation Act (Chapter 161, Florida

Statutes). State agencies regulating the activities are the Florida DER and the Florida Department of Natural Resources (DNR). Federal laws regulating dredge and fill include the River and Harbor Act of 1899, the Clean Water Act of 1977, and the Marine Protection Research and Sanctuaries Act of 1972. The U.S. Army Corps of Engineers is the federal regulatory agency, while the U.S. Fish and Wildlife Service is another concerned agency. Florida DER and DNR and the Army Corps of Engineers have a joint dredge and fill application to facilitate the permit procedure.

Several local communities in Florida have additional regulations against alteration, cutting, pruning, removal, or destruction of mangrove wetlands. In order to obtain the necessary information on application procedures, regulations, and guidelines, contact local county or city planning and building departments, the Florida Department of Environmental Regulation, or the U.S. Army Corps of Engineers (appendix B). These agencies will assist you in identifying the type and scope of information needed to process your application. There are 2 types of permit applications: short form and regular form, depending upon the size and scope of the project. It should be noted that permit review and approval is more closely scrutinized if a project is located adjacent to Class I and Class II waters, state aquatic preserves, or "Outstanding Florida Waters."

Local government comprehensive plans

In Florida the Local Government Comprehensive Planning Act of 1975 (Chapter 163, Florida Statutes) mandates that all local governments prepare, adopt, and implement a comprehensive plan that addresses present and future community growth and development needs. The act requires that each unit of local government (county, city, municipality, town, or village) establish a planning process and prepare, adopt, and implement a comprehensive plan. The law requires that the planning process be ongoing, based on effective public participation, and include regular plan review, update, and appraisal. Every community's plan must contain the following required elements: future land-use plan; traffic circulation; sanitary sewer, solid waste, drainage, and potable water supply; conservation; coastal zone (along the coast); recreation and open space; housing; utilities; and intergovernmental relations. For communities in excess of 50,000, 2 additional elements are required: mass transit and port, aviation, and related facilities. The law requires that local governments set forth principles and standards to guide future development. Unfortunately, the law itself does not provide for any minimum standards or enforcement procedures, particularly in regard to the coastal zone.

Almost all of the counties and municipalities in Florida have adopted comprehensive plans. Many of these local plans are quite detailed with specific provisions for the use of beaches, dunes, marshes, bays, barrier islands, and coastal wetlands.

If you plan to buy property or build in the coastal areas of barrier islands, we suggest that you contact your local planning, zoning, and building departments. Their offices are generally located in or near the county courthouse or the city hall. You need to obtain the necessary zoning, subdivision, or building permits from the local government. These agencies will assist you in determining which kinds of development activities are permitted and which are not. The early contact with the local planning or building departments will enable you to find out which, if any, state and federal permits will be necessary for your project.

Coastal construction permits

The low-lying barrier beaches and coastlines of Florida render coastal construction a matter of particular concern to the state and to local communities. The Florida legislature addressed this concern by enacting the Beach and Shore Preservation Act (Chapter 161, Florida Statutes). The requirements and constraints of this law are in addition to those dealing with water-quality control and dredging and filling laws explained earlier. In 1971 the legislature established a goal that stated in part, "The Legislature finds and declares that the beaches of the State, by their nature, are subject to frequent and severe fluctuations and represent one of Florida's most valuable natural resources and that it is in the public interest to preserve and protect them from imprudent construction which can jeopardize the stability of beach-dune system . . . ."

The legislature directed the Florida Department of Natural Resources, Division of Beaches and Shores, to establish Coastal Construction Setback Lines (CCSBL) that were replaced by Coastal Construction Control Lines (CCCL) in 1978. The CCCL is established under legislative mandate, "so as to define that portion of the beach-dune system which is subject to severe fluctuations based on a 100-year storm surge or other predictable weather conditions, and so as to define the area within which special structural design consideration is required to insure protection of beach-dune system, any proposed structure and adjacent properties, rather than to define a seaward limit for upland structure."

The intent of the law is to regulate coastal construction, to establish coastal construction control lines along the sandy beaches, seaward of which construction may not occur without an authorized permit from the DNR, and to administer a beach erosion control grants-in-aid program. The Florida DNR has established a coastal construction control line for each coastal county with sandy beaches fronting the Atlantic or Gulf after conducting a comprehensive study of the areas' coastal resources and processes (see fig 7.1). Coastal construction is undertaken pursuant to a CCCL permit. The 2 separate permit programs are described in the next few paragraphs.

Coastal Construction Permits (Chapter 161.041, Florida Statutes). The DNR coastal construction permits are required for any construction or change of existing structures and construction or physical activity undertaken for shore protection or erosion control purposes if that activity is located below the mean high-water line of any tidal water of the state.

Specifically included are such structures as dune walkovers, groins, jetties, moles, breakwaters, seawalls, bulkheads, and revetments. Physical activities, such as artificial beach nourishment, inlet sediment bypassing, excavation or maintenance dredging on

Fig. 7.1. This Pinellas County condo was built where the surf zone used to be! The developer violated the spirit of the construction setback law.

inlet channels, and deposition or removal of beach material also require a permit.

   Coastal Construction Control Line Permit (Chapter 161.053, Florida Statutes). If any structure is located seaward of the CCCL (or CCSBL) on any sandy shoreline fronting the Gulf of Mexico or the Atlantic Ocean, permits are required from the DNR before undertaking any alteration. Among the activities covered are ex-

cavation and construction of any dwelling house, hotel, motel, apartment, condominium, seawall, revetment, pool, patio, garage, parking lot, minor structure, and dune restoration. Driving of motorized vehicles or removal of sea oats on the beaches and dunes seaward of the control line also is prohibited except in 3 Atlantic coastal counties. There are certain counties without sandy beaches, and certain types of structures that are exempt from the law. In order to determine if you need any DNR permit, contact the Division of Beaches and Shores, Department of Natural Resources (appendix B).

In 2 cases (Lee and Pinellas counties) the DNR has delegated authority to the coastal counties and municipalities to administer the CCCL requirement. If you plan to buy or build on the oceanfront it is advisable to check with the county or municipality if your property is located seaward of the CCCL because these areas are extremely hazardous, often unsuitable for permanent structures, and create serious environmental and economic problems for property owners as well as taxpayers.

State Land Lease Permit (Chapter 253, Florida Statutes). If your project involves use of sovereign (state-owned) submerged lands in Florida, you need to obtain appropriate permits from the Bureau of State Lands Management of the Florida Department of Natural Resources (appendix B) pursuant to Chapter 253.77 of the Florida Statutes. Sovereignty lands include tidal lands, islands, sand bars, and lands under navigable waters, whether fresh- or saltwater, which Florida gained title to when it became a state. Generally, permits are more closely scrutinized if the project is in a state aquatic preserve, manatee sanctuary, or other environmentally designated sensitive areas. The rules and regulations for the issuance of permits are contained in the Florida Administrative Code Rule 160–21 (Sovereignty Submerged Lands Management Rule), Rule 160–18 (Biscayne Bay Aquatic Preserves Rule), and Rule 160–21 (Florida Aquatic Preserve Rules).

The most common activities included under this permit program are docks for private and commercial use, reclamation of land lost by erosion or avulsion, and dredging of navigation channels. Certain noncommercial activities that are not located within state aquatic preserves or specially designated manatee areas are exempt from any requirements to make application for consent of use. In general, if you suspect that your proposed construction activity might involve use of state-owned lands, we recommend that you contact the Bureau of State Lands Management to determine whether or not you need a permit from the bureau or other state agencies.

Building codes

For residential dwellings other than mobile homes, Florida requires all communities to adopt 1 of 5 acceptable building codes (Chapter 553, Part VI, Florida Statutes), including the One and Two Family Dwelling Code (after the CABO, Council of American Building Officials), the Standard Building Code, the South Florida Building Code, the National Building Code, and the EPCOT Building Code. Of these, only the Standard Building Code is in common use along the West Florida coast. The other codes are used in some east coast counties and inland.

As an example, the Standard Building Code (formerly the Southern Standard Building Code; reference 75, appendix C) was compiled by knowledgeable engineers, architects, and code enforcement officials to regulate the design and construction of buildings and the quality of building materials. The code does have certain hurricane resistance requirements, such as continuity, stability, and anchorage, all related to calculated reference wind speed as modified by height above ground and building shape factors to determine design load.

It is emphasized that the purpose of these codes is to provide *minimum* standards to safeguard lives, health, and property. Communities have the right to strengthen the adopted code in order to improve it or to make it more stringent. By law, such improvements in a code cannot discriminate against materials, products, or construction techniques of proven capabilities; and there must be some unique physiographic condition (for example, geographic type or location, topographic features or absence thereof) that warrants the more stringent requirements. All barrier islands and beaches facing the open ocean and the threat of hurricanes should meet the latter requirement. As a result, numerous communities do have specifications that go beyond the Standard Building Code. The Florida Department of Natural Resources has made recommendations for a Coastal Construction Building Code (reference 75, appendix C) to supplement existing code. Check with your local building inspector to determine the specific code for your area.

Individuals can and should insist on designs and materials that go beyond the *minimum* code requirements (see chapter 6 on construction). Sanibel Island has adopted one of the better codes in Florida with respect to coastal construction. You might contact Sanibel's building inspector's office for a copy of their code to see examples of performance-oriented criteria and coastal issues treated in greater depth. The State Division of Beaches and Shores also provides coastal construction guidelines and may be consulted for advice.

Persons concerned with planning and improving existing building codes should contact the Department of Community Affairs (appendix B). Their Bureau of Emergency management offers the free publication "Hazard Mitigation through Building Codes."

Mobile home regulations

Mobile homes differ in construction and anchorage from "permanent" structures. The design, shape, lightweight construction materials, and other characteristics required for mobility, or for staying within axle-weight limits, create a unique set of potential problems for residents of these dwellings. Because of their thinner walls, for example, mobile homes are more vulnerable to wind and wind-borne projectiles.

Some coastal states have code requirements for mobile homes that are specific for units locating in hurricane-prone areas. Florida does not, although mobile home construction must meet national code requirements. Regulation is through the Department of Highway Safety and Motor Vehicles.

Mobile home anchorage tiedowns are required throughout Florida. Tiedowns make the structure more stable against wind stress (for recommendations, see the section on mobile homes in chapter 6). Older metal tiedowns may be weakened through corrosion, or violations of anchorage or foundation regulations may go undetected unless there are a sufficient number of conscientious inspectors to monitor trailers. One poorly anchored mobile home can wreak havoc with adjacent homes whose owners abided by sound construction practice. Some mobile home park operators or managers are alert to such problems and see that they are corrected; others simply collect the rent.

The spacing of mobile homes should be regulated by local ordinance. Providing residents with open space between homes, this type of ordinance preserves some aesthetic value for a neighborhood. It also helps to maintain a healthier environment. For example, if mobile home septic tanks are closely spaced, there is the potential for groundwater or surface water pollution. Similarly, if mobile homes are built too closely to finger canals, canal water may become polluted.

Prefabricated structure regulation

Modular unit construction is one of the new approaches to construction of multiple-dwelling structures in the coastal zone (see the section on modular unit construction in chapter 6). These prefabricated units are assembled at the shore as multiplexes and condominiums, commonly 2 to 4 stories in height. The Department of Community Affairs sets the building code standards for all prefabricated buildings, including modular unit dwellings. The department follows standard requirements that include wind design speeds of up to 130 mph in some coastal areas.

All such structures must meet the inspection of an independent, third-party testing architect or engineer who signs the structure off if it meets state requirements. The Department of Community Affairs not only approves the plans for such structures but assigns an insignia when the inspection is approved. It is the local building inspector's responsibility, however, to check for the state insignia as well as making sure that service hookups for the buildings meet the local building code.

Historic and archeologic sites

It is the public policy of the state of Florida to protect and preserve historic sites and properties, artifacts, treasure troves, fossil deposits, prehistoric Indian habitations, and objects of antiquity that have historical values or are of interest to the public. The Florida Archives and History Act of 1966 provides the authority to implement these laws through the Division of Archives, History and Records Management in the Florida Department of State (appendix B).

Barrier beaches and coastal areas of Florida have been sites of historic exploration and early native Indian settlements. If you find that there are objects, sites, or structures of some historic archeologic or architectural value on your property, contact the Bureau of Historic Sites for technical assistance to preserve and protect them.

Endangered fish and wildlife species

Florida's environment is blessed with diverse ecosystems that provide habitats for hundreds of common, endangered, and threatened species of fish and wildlife. The protection of wildlife species and habitat is administered by 2 state agencies: the Game and Fresh Water Fish Commission (GFWFC) and the Department of Natural Resources. The state's enabling laws are the Florida Panther Act of 1978, Florida Manatee Sanctuary Act of 1979, and Feeding of Alligator and Crocodile Act. These state laws provide the protection of particular species of wildlife. Additional protection to endangered and threatened fish and wildlife species is provided by the Federal Endangered Species Act of 1973 and Marine Mammal Protection Act of 1972. The U.S. Fish and Wildlife Service of the U.S. Department of the Interior protects the species listed in the law and regulations.

Barrier islands and beaches are habitats for many endangered and threatened fish and wildlife species. The Florida GFWFC and U.S. Fish and Wildlife Service classify endangered and threatened species into three categories:

Endangered. A species, subspecies, or isolated population that is, or soon may be, in immediate danger of extinction unless the species habitat is fully protected and managed for its survival. The American bald eagle, Atlantic green turtle, Florida panther, and West Indian manatee are some examples.

Threatened. A species, subspecies, or isolated population that is very likely to become endangered soon unless the species or its habitat is fully protected and managed for survival. Some examples of this group are loggerhead sea turtle, eastern indigo snake, brown pelican, and mangrove fox squirrel.

Species of special concern. A species, subspecies, or isolated population that warrants special protection because (1) it may,

due to pending environmental degradation or human disturbance, become threatened unless protective management strategies are employed, (2) its status cannot be classified as threatened until more information is available, (3) it occupies such an essential ecological position that its decline might adversely affect associated species, or (4) it has not recovered sufficiently from a past decline or disturbance. Examples in this category include the American alligator, gopher tortoise, roseate spoonbill, limpkin, pine barrens, treefrog, and beach cotton mouse.

The Florida Game and Fresh Water Fish Commission and the U.S. Fish and Wildlife Service (appendix B) provide technical assistance and coordination to identify endangered and threatened species via the development permit review process. Appendix B provides the legal status of endangered and potentially endangered species in Florida as of July 1982.

Other regulations. In addition to the statutes and regulatory requirements outlined above, the attention of state law also has focused on defining ownership of coastal-zone lands (uplands, tidelands, submerged lands). Title to coastal lands is always more complex than for inland property because of the rapid changes due to submergence, erosion, accretion, or shifting of waterways. When purchasing coastal property, make sure you know exactly what you are gaining title to in terms of private ownership.

Florida's coastal future: more regulation

A drive along Florida's Panhandle and Gulf coasts will reveal to even the casual observer that federal, state, and local regulations

Fig. 7.2. A building located too close to the beach on Manasota Key is suffering the inevitable. Rock scattered about the beach reduces its recreational value. Photo by Dinesh Sharma.

have not stopped development in coastal high-hazard zones. Such statutes have not halted the addition of engineering structures that ultimately destroy natural protective beach dune systems.

Present and future citizens of coastal communities should *not* assume that existing statutes and ordinances will guarantee their

safety, that of their property, or protect existing beaches and dunes. Existing regulations address but do not solve the problems of living in the coastal zone (see fiig 7.2).

Most regulations do not prohibit development in high-hazard zones; they only set limits. The national Coastal Barrier Resources Act may deny flood insurance and federal funds for development of certain barrier islands, but it does not necessarily prohibit such development (see fiig 7.3). The National Flood Insurance Program establishes a minimum standard for structural elevation in a flood zone to qualify for flood insurance, but generally it does not prohibit locating in such a flood zone. Similarly, state and local building codes set minimum standards.

Florida's Beach and Shore Preservation Act is an innovative law, establishing the Coastal Construction Control Line (CCCL); but this line is one of the state permitting jurisdiction, not prohibition. Special structural design may be required for proposed structures beyond the line to protect the structure from hazards and to protect adjacent properties as well as the beach dune system. Variances are given, and therein lies the problem. The delicate balance of providing for beach and dune protection while assuring the reasonable use of private property is difficult to achieve. "Reasonable" may be viewed differently between the property owner and the state; but when construction does take place seaward of the control line the impact is likely to be that of the pre-CCCL past, that is, ultimate loss of beach and dunes.

Most of the 250 miles of beaches in Florida that are experiencing "critical erosion" had structures placed too close to the water. As a result, many coastal communities in the state are seeking state and federally subsidized beach nourishment projects to protect the threatened properties. These "protective" projects are extremely costly to general taxpayers as well as to property owners, and are only *temporary* in nature. An effective CCCL program should limit the growth of such problems in the future, given *strict* interpretation and enforcement. Keep in mind also that there are no setback requirements for any structures located on nonsandy shorelines of the tidal areas. As explained in chapter 2, such areas also are experiencing significant erosion due to the sea-level rise and drowning of the coastline (see fig 7.4).

Likewise, the Local Government Comprehensive Planning Act may require planning by local governments, but if this planning takes place without minimum standards or within a system void of land-use planning and zoning that is sensitive to fragile coastal environments and high-energy coastal processes, then it provides no security for the coastal citizen. Even the best of plans is meaningless if there are no provisions and resources for enforcement.

In some respects the approach to coastal regulation is like that of traffic control at a dangerous intersection. The general rules of the road (not to develop in a hazardous zone) are ignored by some drivers, creating unsafe conditions for all (loss of beaches and dunes in front of formerly low-hazard zones). Accidents (property losses) result. Automobile insurance (national flood insurance) spreads the cost to all drivers, recovers a portion of the loss to the victims, but does not make a broken leg whole again or take away paralysis. A group of concerned citizens requests a stoplight (pro-

Fig. 7.3. Development crowding a narrow barrier island, Naples, Florida. Photo by Al Hine.

Fig. 7.4. A large building on Lido Key with a superb sea view. The view will likely become better and better as shoreline retreat continues, but in the next big storm. . . . Photo by Dinesh Sharma.

hibition of unsafe development), but another group believes that a stoplight will back up traffic, slow things down, hurt business (cool the hot economic climate of coastal development). A compromise is reached—the speed limit is reduced (first set of ordinances), but accidents continue to occur (growing property/beach loss), so a blinker light is installed (additional statutes and ordinances). The rules of the road are still being violated (and nature is still gnawing at the shore). Finally a fatal accident occurs; perhaps someone whom all of the community knew and loved is killed (a hurricane strikes, wiping out a community with loss of lives as well as property). Everyone cries "there should have been a stoplight at that intersection five years ago" (additional legislation is still needed). Perhaps 1 or 2 years will go by while surveys are done; and perhaps another fatal accident or accidents will occur. Eventually there will be a stoplight, but the dead will remain dead.

Although Florida has pioneered the enactment of far-reaching coastal legislation, problems remain. The growth in both tourism and coastal development will increase the likelihood of possible conflict between public access rights and private littoral rights. Florida does not have a mechanism for guaranteeing public beach access, although below the high-water line is state land and in the public domain. New laws or changes in existing regulations should be expected. Similarly, building codes, setback lines, and other minimum protective regulations are likely to be more strictly enforced and stiffened as we experience loss of life, property, and shoreline due to imprudent development. In 1983 a barrier island building code was introduced into the Florida legislature. Although the bill was not passed, similar legislation may be forthcoming. Coastal dwellers should expect future changes in laws and regulations, just as they should expect future changes in the beaches and dunes.

Appendix A. Hurricane checklist

Keep this checklist handy for protection of family and property.

When a hurricane threatens

___ Listen for official weather reports.
___ Read your newspaper and listen to radio and television for official announcements.
___ Note the address of the nearest emergency shelter.
___ Know the official evacuation route in advance.
___ Pregnant women, the ill, and the infirm should call a physician for advice.
___ Be prepared to turn off gas, water, and electricity where it enters your home.
___ Fill tubs and containers with water (one quart per person per day).
___ Make sure your car's gas tank is full.
___ Secure your boat. Use long lines to allow for rising water.
___ Secure movable objects on your property:

  ___ doors and gates
  ___ outdoor furniture
  ___ garden tools hoses
  ___ garbage cans
  ___ bicycles or large sports equipment
  ___ barbecues or grills

___ Board up or tape windows and glassed areas. Close storm shutters. Draw drapes and window blinds across windows and glass doors. Remove furniture in their vicinity.

___ Stock adequate supplies:

  ___ transistor radio
  ___ fresh batteries
  ___ canned heat
  ___ hammer
  ___ boards
  ___ pliers
  ___ hunting knife
  ___ tape
  ___ first-aid kit
  ___ prescribed medicines
  ___ water purification tablets
  ___ insect repellent
  ___ gum, candy
  ___ life jackets
  ___ charcoal bucket and charcoal
  ___ buckets of sand
  ___ flashlights
  ___ candles
  ___ matches
  ___ nails
  ___ screwdriver
  ___ ax*
  ___ rope*
  ___ plastic drop cloths, waterproof bags, ties
  ___ containers for water
  ___ disinfectant
  ___ canned food, juices, soft drinks (see below)
  ___ hard-top headgear
  ___ fire extinguisher
  ___ can opener and utensils

___ Check mobile-home tiedowns.

*Take an ax (to cut an emergency escape opening) if you go to the upper floors or attic of your home. Take rope for escape to ground when water subsides.

Suggested storm food stock for family of four

— two 13-oz. cans evaporated milk
— four 7-oz. cans fruit juice
— 2 cans tuna, sardines, Spam, chicken
— three 10-oz. cans vegetable soup
— 1 small can of cocoa or Ovaltine
— one 15-oz. box raisins or prunes
— salt
— pet food?
— one 14-oz. can cream of wheat or oatmeal
— one 8-oz. jar peanut butter or cheese spread
— two 16-oz. cans pork and beans
— one 2-oz. jar instant coffee or tea
— 2 packages of crackers
— 2 pounds of sugar
— 2 quarts of water per person

Special precautions for apartments/condominiums

— Make one person the building captain to supervise storm preparation.
— Know your exits.
— Count stairs on exits; you'll be evacuating in darkness.
— Locate safest areas for occupants to congregate.
— Close, lock, and tape windows.
— Remove loose items from terraces (and from your absent neighbors' terraces).
— Remove or tie down loose objects from balconies or porches.

— Assume other trapped people may wish to use the building for shelter.

Special precautions for mobile homes

— Pack breakables in padded cartons and place on floor.
— Remove bulbs, lamps, mirrors—put them in the bathtub.
— Tape windows.
— Turn off water, propane gas, electricity.
— Disconnect sewer and water lines.
— Remove awnings.
— **Leave**.

Special precautions for businesses

— Take photos of building and merchandise.
— Assemble insurance policies.
— Move merchandise away from plate glass.
— Move merchandise to as high a location as possible.
— Cover merchandise with tarps or plastic.
— Remove outside display racks and loose signs.
— Take out lower file drawers, wrap in trash bags, and store high.
— Sandbag spaces that may leak.
— Take special precautions with reactive or toxic chemicals.

If you remain at home

— Never remain in a mobile home; seek official shelter.
— Stay indoors. Remember, the first calm may be the hurricane's eye. Remain indoors until an official all-clear is given.

___ Stay on the "downwind" side of the house. Change your position as the wind changes.

___ If your house has an "inside" room, it may be the most secure part of the structure.

___ Keep continuous communications watch for *official* information on radio and television.

___ Keep calm. Your ability to meet emergencies will help others.

If evacuation is advised

___ Leave as soon as you can. Follow official instructions only.

___ Follow official evacuation routes unless those in authority direct you to do otherwise.

___ Take these supplies:

    ___ change of warm, protective clothes

    ___ first-aid kit

    ___ baby formula

    ___ identification tags: include name, address, and next of kin (wear them)

    ___ flashlight

    ___ food, water, gum, candy

    ___ rope, hunting knife

    ___ waterproof bags and ties

    ___ can opener and utensils

    ___ disposable diapers

    ___ special medicine

    ___ blankets and pillows in waterproof casings

    ___ radio

    ___ fresh batteries

    ___ bottled water

    ___ purse, wallet, valuables

    ___ life jackets

    ___ games and amusements for children

___ Disconnect all electric appliances except refrigerator and freezer. Their controls should be turned to the coldest setting and the doors kept closed.

___ Leave food and water for pets. Seeing-eye dogs are the only animals allowed in the shelters.

___ Shut off water at the main valve (where it enters your home).

___ Lock windows and doors.

___ Keep important papers with you:

    ___ driver's license and other identification

    ___ insurance policies

    ___ property inventory

    ___ Medic Alert or other device to convey special medical information

During the hurricane

___ Stay indoors and away from windows and glassed areas.

___ If you are advised to evacuate, **do so at once**.

___ Listen for continuing weather bulletins and official reports.

___ Use your telephone only in an emergency.

___ Follow official instructions only. Ignore rumors.

— Keep **open** a window or door on the side of the house opposite the storm winds.
— Beware of the **"eye of the hurricane."** A lull in the winds does not necessarily mean that the storm has passed. Remain indoors unless emergency repairs are necessary. Exercise caution. Winds may resume suddenly, in the opposite direction and with greater force than before. Remember, if wind direction does change, the open window or door must be changed accordingly.
— Be alert for rising water.
— If electric service is interrupted, note the time.
 — Turn off major appliances, especially air conditioners.
 — Do not disconnect refrigerators or freezers. Their controls should be turned to the coldest setting and doors closed to preserve food as long as possible.
 — Keep away from fallen wires. Report location of such wires to the utility company.
— If you detect **gas**:
 — Do not light matches or turn on electrical equipment.
 — Extinguish all flames.
 — Shut off gas supply at the meter.*
 — Report gas service interruptions to the gas company.
— **Water**:
 — The only **safe** water is the water you stored before it had a chance to come in contact with flood waters.

*Gas should be turned back on only by a gas serviceman or licensed plumber.

— Should you require an additional supply, be sure to boil water for 30 minutes before use.
— If you are unable to boil water, treat water you will need with water purification tablets.
Note: An official announcement will proclaim tap water "safe." Treat all water except stored water until you hear the announcement.

After the hurricane has passed

— Listen for official word of danger having passed.
— Watch out for loose or hanging power lines as well as gas leaks. People have survived storms only to be electrocuted or burned. Fire protection may be nil because of broken power lines.
— Walk or drive carefully through the storm-damaged area. Streets will be dangerous because of debris, undermining by washout, and weakened bridges. Watch out for poisonous snakes and insects driven out by flood waters.
— Eat nothing and drink nothing that has been touched by flood waters.
— Place spoiled food in plastic bags and tie securely.
— Dispose of all mattresses, pillows, and cushions that have been in flood waters.
— Contact relatives as soon as possible.

Note: If you are stranded, signal for help by waving a flashlight at night or white cloth during the day.

Appendix B. A guide to local, state, and federal agencies involved in coastal development

Numerous agencies at all levels of government are engaged in planning, regulating, permitting, or studying coastal development and resources in Florida. These agencies provide information on development to the homeowner, developer, or planner and issue permits for various phases of construction and information on particular topics.

Aerial photography, coastal construction control line maps, ortho-photo maps, and remote-sensing imagery

> State Topographic–Aerial Survey Engineer
> Florida Department of Transportation
> Haydon Burns Building
> 605 Suwanee Street
> Tallahassee, FL 32301
> (904) 488–2250
>
> Division of Beaches and Shores
> Florida Department of Natural Resources
> 3900 Commonwealth Boulevard
> Tallahassee, FL 32303
> (904) 488–3180

> U.S. Geological Survey
> 325 John Knox Road
> Tallahassee, FL 32301
> (904) 385–7145
>
> Coastal Zone Studies
> Department of Political Science
> University of West Florida
> Pensacola, FL 32504
> (904) 476–9500
>
> Local county or municipal governments
> Attn: planning, zoning, or building departments.

Beach erosion

Information on barrier beach erosion, inlet migration, and erosion control alternatives is available from the following agencies:

> Division of Beaches and Shores
> Florida Department of Natural Resources
> 3900 Commonwealth Boulevard
> Tallahassee, FL 32303
> (904) 488–3180

Coastal Archives Library
Department of Coastal and Oceanographic Engineering
Weil Hall
University of Florida
Gainesville, FL 32611
(904) 392-2710

Coastal Zone Studies
Department of Political Science
University of West Florida
Pensacola, FL 32504
(904) 476-9500

U.S. Army Corps of Engineers
Jacksonville District Office
P.O. Box 4970
400 East Bay Street
Jacksonville, FL 32232

Coastal Engineering Specialist
Florida Sea Grant-Marine Advisory Program
G022 McCarty Hall
University of Florida
Gainesville, FL 32611
(904) 392-2460

Local county or municipal governments
Attn: planning or engineering departments.

Bridges and causeways

The U.S. Coast Guard has jurisdiction over issuing permits to build bridges or causeways that will affect navigable waters. Information for peninsular Florida from Fernandina Beach to Panama City:

Commander, 7th Coast Guard District
1018 Federal Building
51 First Avenue, S.W.
Miami, FL 33130
(305) 350-4108

For areas west of Panama City, contact:

Commander, 8th Coast Guard District
500 Camp Street
New Orleans, LA 70130
(504) 589-6298

Building codes, planning, and zoning

Most communities have adopted comprehensive plans and building codes under the Southern Standard Building Code and in some cases under the improved South Florida Building Code. Check with your county or city building department for permitted uses and building codes. The existing building codes in Florida do not protect the structures from hurricane, flood, wind, and erosion damage. If you intend to build on barrier islands, we advise you to obtain an excellent guide from Texas as well as the Florida codes.

Model Minimum Hurricane Resistant
Building Codes for the Texas Gulf Coast
Texas Coastal and Marine Council
P.O. Box 13407
Austin, TX 78711
(512) 475-5849

Coastal Construction Building Code Guidelines
Division of Beaches and Shores
Florida Department of Natural Resources
3900 Commonwealth Boulevard
Tallahassee, FL 32303
(904) 488-3180

Coastal zone planning and management program

Florida adopted the Coastal Management Program (CMP) pursuant to the Florida Coastal Zone Management Act of 1978 (Chapter 380, Florida Statutes) and the Federal Coastal Zone Management Act of 1972. Florida's CMP did not create any new agency but provides for coordination and consistency in the implementation of various federal and state programs affecting coastal areas and barrier islands. For information on the CMP and designated barrier islands, contact:

Office of Coastal Management
Department of Environmental Regulation
2600 Blair Stone Road
Tallahassee, FL 32301
(904) 488-4805

Dredging, filling, and construction in coastal waters

Florida laws require that all those who wish to dredge, fill, or otherwise alter wetlands, marshes, estuarine bottoms, or tidelands apply for a permit from the appropriate state, federal, and local governments. For information, write or call the following agencies.

For the standard state permit on dredging and filling, contact:

Florida Department of Environmental Regulation
Twin Towers Building
2600 Blair Stone Road
Tallahassee, FL 32301
(904) 488-0130

For short-form dredge and fill application permits, contact appropriate district offices of the Department of Environmental Regulation.

For erosion control structures and coastal construction control line permits, write or call:

Division of Beaches and Shores
Florida Department of Natural Resources
3900 Commonwealth Boulevard
Tallahassee, FL 32303
(904) 488-3180

Easements and submerged land leases for docks, piers, etc., must be obtained from:

Bureau of State Lands Management
Florida Department of Natural Resources
3900 Commonwealth Boulevard
Tallahassee, FL 32303
(904) 488-9120

Federal law requires that any person who wishes to dredge, fill, or place any structure in navigable water (almost any body of water) apply for a permit from the U.S. Army Corps of Engineers.

Permit Branch
U.S. Army Corps of Engineers
P.O. Box 4970
400 East Bay Street
Jacksonville, FL 32232
(904) 791-2887

The Army Corps of Engineers has 10 additional field offices in Florida. Consult your area telephone directory for the U.S. Government listing in the white pages.

Dune alteration and vegetation removal

Florida laws prohibit the destruction, damaging, or removal of sea grasses, sea oats, or sand dunes and berms. Individual counties or cities might have ordinances pertaining to dune alteration and vegetation removal. Permits for certain work and alteration may be obtained from local county or city planning and building departments. For permits to clear or alter dunes or beaches seaward of the coastal construction control lines, write or call

Division of Beaches and Shores
Department of Natural Resources
3900 Commonwealth Boulevard
Tallahassee, FL 32303
(904) 488-3180

Health, sanitation, and water quality

County health departments are in charge of issuing on-site septic system permits for sewage treatment plants with a capacity of less than 2,000 gallons per day. For the necessary information and permit process, contact your local health department or:

Division of Environmental Services
Department of Health and Rehabilitative Services
1317 Winewood Boulevard
Tallahassee, FL 32301
(904) 488-4070

For sewage systems with a capacity larger than 2,000 gallons per day, the Florida Department of Environmental Regulation issues the permits. (For address, see Dredging and filling listing.)

History and archeology

If you suspect that your property may have an archeologic or historic site, write or call:

Bureau of Historic Sites and Properties
Division of Archives, History and Records Management
Florida Department of State
R.A. Gray Building
Tallahassee, FL 32301
(904) 488-1480, 488-2333

Hurricane information and weather

The National Oceanic and Atmospheric Administration is the best agency from which to request information on hurricanes. NOAA stormflood evacuation maps are prepared for vulnerable coastal areas and cost $2.00 each. For details, call or write:

Distribution Division (C-44)
National Ocean Survey
NOAA
Riverdale, MD 20840
(301) 463-6990

For hurricane probability charts for your area as well as weather and hurricane warning information, contact:

National Hurricane Center
1320 South Dixie Highway
Coral Gables, FL 33146
(305) 665-0413

Hurricane evacuation and disaster assistance

Contact your local county or city Civil Defense or Disaster Preparedness office for hurricane evacuation and hurricane shelter information. Local radio and TV stations provide hurricane warning and evacuation bulletins when storms threaten an area. For information on hurricane disaster response, recovery, and assistance, contact:

Disaster Response and Recovery Assistance
Bureau of Disaster Preparedness
Florida Department of Community Affairs
2571 Executive Center Circle, East
Tallahassee, FL 32301
(904) 488-1900
(904) 488-1320 (in case of emergency)

Federal Disaster Assistance Administration
Region 4 office
Suite 750
1375 Peachtree Street, N.E.
Atlanta, GA 30309

Insurance

In coastal areas special building requirements must often be met to obtain flood or wind storm insurance. To find out the requirements in your area, check with your local building department and insurance agent. Further information is available from:

Federal Insurance Administration
National Flood Insurance Program
Federal Emergency Management Agency
Washington, DC 20472
(202) 755-5290

Division of Insurance-Consumer Services
Florida Department of Insurance and Treasurer
The Capitol
Tallahassee, FL 32301
(904) 488-2660

Land planning and land use

Local county and municipal governments have adopted comprehensive land-use plans and zoning and building codes under the state law. It is advisable to contact these agencies, preferably before you buy land on a barrier island or coastal area. For additional information, contact:

Division of Local Resources Management
Florida Department of Community Affairs
2571 Executive Center Circle, East
Tallahassee, FL 32301
(904) 488-2356

Large-scale development projects requiring Development of Regional Impact (DRI) studies must meet approvals by the following state agency:

Bureau of Land and Water Management
Florida Department of Community Affairs
2571 Executive Center Circle, East
Tallahassee, FL 32301
(904) 488-4925

Area Regional Planning Council approvals also are necessary for DRIs before state agency approvals are granted.

Land purchase and sales

When acquiring property or a condominium—whether in a subdivision or not—consider the following: (1) Owners of property next to dredged canals should make sure that the canals are designed for adequate flushing to keep waters from become stagnant. Requests for federal permits to connect extensive canal systems to navigable waters are frequently denied. (2) Descriptions and surveys of land in coastal areas are very complicated. Old titles granting fee-simple rights to property below the high-tide line may not be upheld in court; titles should be reviewed by a competent attorney before they are transferred. A boundary described as the high-water mark may be impossible to determine. (3) Ask about the provision of sewage disposal and utilities including water, electricity, gas, and telephone. (4) Be sure any promises of future improvements, access, utilities, additions, common property rights, etc., are in writing. (5) Be sure to visit the property and inspect it carefully before buying it.

Land preservation

Several barrier beaches are being considered by the state for public acquisitor under 3 state programs: the Environmentally Endangered Lands Program, the Save Our Coast Program, and the Conservation and Recreational Lands Program. If you own large parcels of environmentally sensitive land on barrier islands or coastal areas and prefer to have it preserved for future generations to enjoy, contact one of these agencies.

On the other hand, if you plan to buy barrier island property, it would be advisable to contact the local government agency as well as the following state agencies to determine if there could be development and permitting problems.

Bureau of Land Acquisition
Division of State Lands
Department of Natural Resources
3900 Commonwealth Boulevard
Tallahassee, FL 32301
(904) 488-2725

Inter-Agency Management Committee
Save Our Coast Program
Office of Planning and Budgeting
Executive Office of the Governor
The Capitol
Tallahassee, FL 32301
(904) 488-5551

Land sales—subdivisions

Subdivisions containing more than 50 lots and offered in interstate commerce must be registered with the Office of Interstate Land Sales Registration (as specified by the Interstate Land Sales Full Disclosure Act). Prospective buyers must be provided with a property report. This office also produces a booklet entitled *Get the Facts . . . Before Buying Land* for people who wish to invest in property. Information on subdivision property and land investment is available from:

Office of Interstate Land Sales Registration
U.S. Department of Housing and Urban Development
Washington, DC 20410

Office of Interstate Land Sales Registration
Atlanta Regional Office
U.S. Department of Housing and Urban Development
230 Peachtree Street, N.W.
Atlanta, GA 30303
(404) 525-4364

Soils and septic systems

Soil type is important in terms of (1) the type of vegetation it can support, (2) the type of construction technique it can withstand, (3) its drainage characteristics, and (4) its ability to accommodate septic systems. For detailed information on soil characteristics and limitations and permitting rules for septic systems, contact:

Local county Soil Conservation Service office, U.S. Department of Agriculture (listing in the telephone book)

Soil Conservation Service
U.S. Department of Agriculture
University of Florida
Gainesville, FL 32611

Local county health department

Environmental Services
Department of Health and Rehabilitative Services
1323 Winewood Boulevard
Tallahassee, FL 32301
(904) 488-4070

U.S. Fish and Wildlife Service
Department of the Interior
P.O. Box 2676
Vero Beach, FL 32960
(904) 562-3909

Water supply and pollution control

If your plan involves draining of land or a large water supply system, contact the appropriate area water management district for rules and permit process. Construction of any sewage or solid waste disposal facilities requires permits from the Florida Department of Environmental Regulation. Contact appropriate district office of the DER. (See Dredging and filling listing for addresses.)

Wildlife species and habitat protection

For the conservation and protection of fish and wildlife species and their habitat, contact the office of the Game and Fresh Water Fish Commission in your area. Also contact:

Appendix C. Useful references

Following is a list of publications compiled according to topic. Most of these publications are of a nontechnical nature and are intended for the nonexpert reader. Floridians who wish to delve into the more detailed literature are referred to the outstanding library of the Department of Coastal Engineering at the University of Florida in Gainesville. This may well be one of the world's greatest repositories of coastal literature. Other important sources include the U.S. Army Corps of Engineers' district offices in Jacksonville and Mobile, offices of the Federal Emergency Management Agency, and various state agency offices in Tallahassee concerned with marine affairs.

General

1. Living with the Shore series, Orrin H. Pilkey, Jr., and William J. Neal, eds. This series addresses problems and concerns of coastal living and management for all Atlantic, Gulf coast, and Pacific coast states, as well for Lake Erie and Lake Michigan. Books for North Carolina, Texas, South Carolina, Long Island's south shore, Louisiana, and Florida's east coast are now available to the public. The present volume is the seventh book in the series, published by Duke University Press, 6697 College Station, Durham, NC 27708. Paperback and hardcover copies may be ordered from Duke Press for $9.75 and $22.75, respectively.

2. *The Beaches Are Moving: The Drowning of America's Shoreline*, by Wallace Kaufman and Orrin H. Pilkey, Jr., 1979, 1983. This highly readable account of the state of America's coastline explains natural processes at work at the beach, provides a historical perspective of man's relation to the shore, and offers practical advice on how to live in harmony with coastal environments. Originally published by Anchor Press/ Doubleday, it is now available in paperback ($9.75) from Duke University Press, 6697 College Station, Durham, NC 27708.

3. *Coastal Design: A Guide for Builders, Planners, and Homeowners*, by Orrin H. Pilkey, Sr., Walter D. Pilkey, Orrin H. Pilkey, Jr., and William J. Neal, 1983. The "umbrella" book for the Living with the Shore series, this volume emphasizes principles of shoreline construction and is intended to be a companion volume to the individual state books of the series. Van Nostrand Reinhold, 135 West 50th Street, New York, NY 10020, 224 pp., $25.50.

4. *Terrigenous Clastic Depositional Environments*, Miles Hayes and Tim Kana, eds., 1976. Although compiled for a professional field course, this text provides an excellent detailed treatment of various sedimentary environments that makes good reading for the interested nonscientist. The numerous

photographs and diagrams support the text descriptions of depositional systems in rivers, dunes, deltas, tidal flats and inlets, salt marshes, barrier islands, and beaches. Most of the examples are from South Carolina. Technical report no. 11-CRD (184 pp.) of the Coastal Research Division, Department of Geology, University of South Carolina, Columbia, SC 29208. Probably easiest to obtain through a college or university library.

5. *Coastal Geomorphology*, D. R. Coates, ed., 1973. A collection of technical papers, including R. Dolan's "Barrier Islands: Natural and Controlled." Interesting reading for anyone willing to overlook or wade through the jargon of coastal scientists. Published by the State University of New York, Binghamton, NY 13901. Available in university libraries.

6. *Barrier Island Handbook*, by Steve Leatherman, 1979. A nontechnical, easy-to-read paperback about barrier island dynamics and coastal hazards. Many of the examples are from the Maryland and New England coasts but are applicable to West Florida as well. Available from Coastal Publications, 5201 Burke Drive, Charlotte, NC 28208 ($5.75).

7. *Barrier Islands from the Gulf of St. Lawrence to the Gulf of Mexico*, Steve Leatherman, ed., 1979. This collection of technical papers presents some of the current geographical research on barrier islands. Of particular relevance to interested West Florida residents is the lead paper by Miles Hayes entitled "Barrier Island Morphology as a Function of Tidal and Wave Regime." Published by Academic Press, the book is available through most college and university libraries.

8. *Beach Processes and Sedimentation*, by Paul Komar, 1976. The most up-to-date textbook on beaches and beach processes, this title is recommended only to serious students of the beach. Published by Prentice-Hall, Englewood Cliffs, NJ 07632, and available through university libraries.

9. *Sea Islands of the South*, by Diana and Bill Gleasner, 1980. An excellent visitor's guide to the southeastern coast of the United States, including naturalist information and descriptions of developed barrier islands. You will find descriptions, explanations, and identifications of everything from dunes to tides to birds and shells; guides to visitor information centers, accommodations, activities, and sightseeing points of interest. This guide is especially handy for the first-time traveler through North Carolina to Florida, but it may be of interest to natives, too. Published by East Woods Press and available in most coastal bookstores.

10. *Coastal Ecosystem Management*, by John Clark, 1977. This 929-page text covers aspects of the coastal zone from descriptions of processes and environments to legal controls and outlines for management programs. Essential reading for planners and beach community managers. Published by John Wiley and Sons, it is available through most university libraries.

11. *Coastal Mapping Handbook*, M. Y. Ellis, ed., 1978. A primer on coastal mapping, this book outlines the various types of

maps, charts, and photography available; it then gives sources and uses for such products, data on state coastal mapping programs, informational appendixes, and examples. It is a valuable starting reference for anyone interested in maps or mapping. For sale by the Superintendent of Documents, U.S. Government Printing Office, Washington, DC 20402 (stock no. 024-001-03046-2), 200 pp.

12. "The Law and the Coast in a Clamshell, Part IV: The Florida Approach," by Peter H. F. Graber, 1981. A concise summary of Florida coastal law and its historic development, including footnotes on significant cases and court decisions, this article appeared in *Shore and Beach* magazine (vol. 49, no. 3, pp. 13-20). *Shore and Beach* is available through most college and university libraries.

13. *Our Changing Coastline*, by F. P. Shepard and H. R. Wanless, 1971. A state-by-state rundown of the recent geologic history of the entire U.S. shoreline. Illustrations are mainly aerial photographs. The book is now a bit out-of-date, but it still furnishes dramatic proof of the highly dynamic nature of our coastlines. McGraw-Hill, 1221 Avenue of the Americas, New York, NY 10020, 579 pp.

14. *The Encyclopedia of Beaches and Coastal Environments*, M. L. Schwartz, ed., 1982. Written for the geologist rather than the layman, this is a very informative and complete volume that discusses everything you would want to know on the subject. Hutchinson-Ross, Stroudsburg, Pa. 18360.

15. *Waves and Beaches*, by Willard Bascom, 1980. This highly readable primer on beach processes is available in paperback from most coastal zone bookstores. Anchor Press / Doubleday.

16. *At the Sea's Edge*, by William T. Fox, 1983. This text is billed as an introduction to coastal oceanography for the amateur naturalist, but it is not easy reading for the non-scientist. Nonetheless, this is one of the best, most up-to-date, and most complete volumes available for those who want to learn about the shoreline and are starting from scratch. Prentice-Hall, Englewood Cliffs, NJ 07632, 317 pp.

17. *Barrier Island Ecology of Cape Lookout National Seashore*, by Paul Godfrey. An excellent summary of how islands in North Carolina have evolved, the book emphasizes the role of overwash and vegetation in island evolution. Despite its North Carolina focus, this volume is pertinent to an understanding of West Florida's barrier islands. National Park Service, Department of the Interior, scientific monograph no. 9.

18. *Florida Coastal and Environmental Information*, by Lucille Lehmann and Todd L. Walton. A useful pamphlet of addresses and sources for all kinds of information important to coastal residents. Available from the Marine Advisory Program, Florida Cooperative Extension Service, University of Florida, Gainesville, FL 32611.

19. *Florida Sea Grant Publication Catalog*. This annual lists a wide variety of useful publications for the coastal dweller. Many of the publications cited are available free of charge. The

catalog itself can be obtained from the Sea Grant Marine Advisory Program, University of Florida, Gainesville, FL 32611.

Films

20. *It's Your Coast.* This is a 28-minute film produced by NOAA on coastal zone management problems. Available from the Marine Advisory Program, University of Florida, Gainesville, FL 32611.

21. *Portrait of a Coast.* This spectacular 29-minute film shows, among other things, a major storm on the Massachusetts coast. It also addresses the interrelated problems of rising sea level, coastal erosion, and shoreline stabilization. This film is very pertinent to the West Florida situation. Available from Circle Oak Productions, 73 Giudle Ridge Drive, Katonah, NY 10536.

22. *Tornadoes.* This 15-minute NOAA film shows how the tornado warning system works as well as showing scenes of tornado damage. Available from the Marine Advisory Program, University of Florida, Gainesville, FL 32611.

23. *Hurricane Before the Storm.* This 29-minute film centers on Hurricane Eloise, which hit northeast Florida, and discusses how to save life and property. Available from the Marine Advisory Program, University of Florida, Gainesville, FL 32611.

Shoreline engineering

24. *Shore Protection Manual*, by the U.S. Army Corps of Engineers, 1973. The "bible" of shoreline engineering, published in 3 volumes at $14.25. Request publication 08-0022-00077 from the Superintendent of Documents, U.S. Government Printing Office, Washington, DC 20402.

25. *Help Yourself*, by the U.S. Army Corps of Engineers. This brochure addressing erosion problems in the Great Lakes region may be of interest to barrier island residents. It outlines shoreline processes and illustrates a variety of shoreline engineering devices used to control erosion. Available free of charge from the U.S. Army Corps of Engineers, North Central Division, 219 South Dearborn Street, Chicago, IL 60604.

26. *Publications List, Coastal Engineering Research Center (CERC) and Beach Erosion Board (BEB)*, by the U.S. Army Corps of Engineers. A list of published research by the Corps of Engineers. Available free of charge from the U.S. Army Corps of Engineers, Coastal Engineering Research Center, Kingman Building, Fort Belvoir, VA 22060.

27. *Low-Cost Shore Protection*, by the U.S. Army Corps of Engineers, 1982. A set of 4 reports written for the layman, this book includes the introductory report, a property owner's guide, a guide for local government officials, and a guide for engineers and contractors. The reports are a summary of the Shoreline Erosion Control Demonstration Program and suggest a wide range of engineering devices and techniques to stabilize shorelines, including beach nourishment and vegetation. In adopting these approaches, one should keep in mind

that they are short-term measures and may have unwanted side effects. The reports are available from John G. Housley, Section 54 Program, U.S. Army Corps of Engineers, USACE (DAEN-CWP-F), Washington, DC 20314.

28. *Shore Management Guidelines* and *Shore Protection Guidelines*, by the U.S. Army Corps of Engineers, 1972. These 2 publications are designed to instruct the public on structural solutions in solving shoreline erosion problems. Both are available from the U.S. Army Corps of Engineers, Department of the Army, Washington, DC 20314.

Hurricanes

29. *Early American Hurricanes, 1492–1870*, by D. M. Ludlum, 1963. An excellent summary of the stormy history of the Atlantic and Gulf coasts that provides a lesson on the frequency, intensity, and destructive potential of hurricanes. Published by the American Meteorological Society (Boston), it is available in many public and university libraries.

30. *Hurricanes and Coastal Storms*, Earl Baker, ed., 1980. These technical papers, presented at a 1979 national conference in Orlando, deal with hurricane and storm awareness, evacuation, and mitigation. They make good reading for planners, developers, and coastal community officials. Available from the Marine Advisory Program, University of Florida, Gainesville, FL 32611.

31. *Bibliography on Hurricanes and Severe Storms of the Coastal Plains Region*, and *Supplement*, by the Central Plains Center for Marine Development Services, 1970 and 1972. A list of references that provides a good starting point for people seeking detailed information on hurricanes and hurricane research. Available through college and university libraries.

32. *Hurricane Information and Gulf Tracking Chart*, by NOAA, 1974. An important brochure that describes hurricane characteristics and lists safety precautions. Available from the Superintendent of Documents, U.S. Government Printing Office, Washington, DC 20402.

33. *Post-Disaster Reports*, by the U.S. Army Corps of Engineers. Available from the Corps' Jacksonville and Mobile district offices, these post-hurricane reports provide valuable documentation of storm damage and are particularly useful for those who wish to prevent recurrences.

Site selection

34. *Ecological Determinants of Coastal Area Management* (2 vols.), by Francis Parker, David Brower, and others, 1976. Volume 1 defines the barrier island and related lagoon-estuary systems and the natural processes that operate within them. It then outlines man's disturbing influences on island environments and suggests management tools and techniques. Volume 2 is a set of appendixes that includes information on coastal ecological systems, man's impact on barrier islands, and tools and techniques for coastal area management. Also

contains a good barrier island bibliography. Available from the Center for Urban and Regional Studies, University of North Carolina, 108 Battle Lane, Chapel Hill, NC 27514.

35. *Coastal Ecosystems: Ecological Considerations for Management of the Coastal Zone*, by John Clark, 1974. A clearly written, well-illustrated book on the applications of ecological principles to major coastal zone environments. Available from the Publications Department, Conservation Foundation, 1717 Massachusetts Avenue, N.W., Washington, DC 20036.

36. *Coastal Environmental Management*, by the Conservation Foundation, 1980. Guidelines for conservation of resources and protection against storm hazards, including ecological descriptions and management suggestions for coastal uplands, floodplains, wetlands, banks and bluffs, dunelands, and beaches. Part 2 presents a complete list of federal agencies and their legal authority to regulate coastal zone activities. A good reference for planners and people interested in good land management. Available from the Superintendent of Documents, U.S. Government Printing Office, Washington, DC 20402.

37. *Natural Hazard Management in Coastal Areas*, by G. F. White and others, 1976. The most recent summary of hazards along the U.S. coast, this volume discusses adjustments to such hazards as well as explaining hazard-related federal policy and programs. Also summarized are hazard management and coastal land planning programs in each state. Ap-

pendixes include a directory of agencies, an annotated bibliography, and information on hurricanes. An invaluable reference, recommended to developers, planners, governmental managers, and concerned coastal residents. Available from the Office of Coastal Zone Management, National Oceanographic and Atmospheric Administration, 3300 Whitehaven Street, N.W., Washington, DC 20235.

38. *Guidelines for Identifying Coastal High Hazard Zones*, by the U.S. Army Corps of Engineers, 1975. This report emphasizes "coastal special flood-hazard areas" (coastal floodplains subject to a 1 percent chance of flooding by hurricane surge in any given year). It provides technical guidelines for conducting uniform flood insurance studies and outlines methods of obtaining 100-year storm-surge elevations. Recommended to island planners. Available from the U.S. Army Corps of Engineers, Galveston District Office, Galveston, TX 77553.

39. *Report of Investigation on the Environmental Effects of Private Water-front Lands*, by W. Barada and W. M. Partington, 1972. An enlightening reference, this study treats the effects of finger canals on water quality. Available from the Environmental Information Center, Florida Conservation Foundation, Inc., 935 Orange Avenue, Winter Park, FL 32789.

40. *Know Your Mud, Sand, and Water: A Practical Guide to Coastal Development*, by K. M. Jurgensen, 1976. Clearly and simply written, this pamphlet describes various island environments and their aptness for development. Recommended

to island dwellers. Available from UNC Sea Grant, North Carolina State University, Box 8065, Raleigh, NC 27695.

Water

41. *Ground Water in the Coastal Plains Region: A Status Report and Handbook*, compiled by A. D. Park, 1979. This report was prepared for the Coastal Plains Regional Commission and addresses the subject of groundwater for 5 southeastern states. Although specific West Florida coastal groundwater problems are not presented, other states' problems and programs provide a basis for comparison. There is an extensive list of references, including studies on saltwater intrusion, aquifer depletion, and water contamination associated with waste disposal. The report is available from the Coastal Plains Regional Commission, 215 East Bay Street, Charleston, SC 29401.

42. *Your Home Septic System: Success or Failure*. This brochure provides answers to commonly asked questions on home septic systems and lists agencies that supply information on septic tank installation and operation. Available from UNC Sea Grant, North Carolina State University, Box 8065, Raleigh, NC 27695.

43. *Water Resources Development*, by the U.S. Army Corps of Engineers. A basic guide of 191 pages. Available from the U.S. Army Corps of Engineers, Jacksonville District Office, Jacksonville, FL 32201.

Vegetation

44. *Mangroves: A Guide for Planting and Maintenance*, by John Stevely and Larry Rabinowitz, 1982. How to identify, plant, and maintain mangroves, and why they are important. Available from Sea Grant Marine Advisory Program, University of Florida, Gainesville, FL 32611.

45. *Dune Restoration and Revegetation Manual*, by Jack Salmon and others, 1982. A manual for both homeowners and community officials anywhere in the southeastern United States. Available from the Sea Grant Marine Advisory Program, University of Florida, Gainesville, FL 32611.

46. *Stabilization of Beaches and Dunes by Vegetation in Florida*, by John Davis. Guidelines for the use of vegetation in Florida nearshore areas. Available from the Sea Grant Marine Advisory Program, University of Florida, Gainesville, FL 32611.

47. *How to Build and Save Beaches and Dunes*, by John A. Jagschitz and Robert C. Wakefield, 1971. This 12-page pamphlet outlines how to build dunes using brush, snow fencing, or American beach grass, with most of the emphasis on the planting and care of beach grass. Available from the Rhode Island Agricultural Experiment Station, Woodward Hall, University of Rhode Island, Kingston, RI 02881.

48. *Building and Stabilizing Coastal Dunes with Vegetation* (UNC–SG–82–05) and *Planting Marsh Grasses for Erosion Control* (UNC–SG–81–09), by S. W. Broome, W.W. Wood-

house, Jr., and E. D. Seneca, 1982. These publications on the use of vegetation as stabilizers are available from Sea Grant, Box 8065, Raleigh, NC 27695. State the publication number with your request.

49. *The Dune Book: How to Plant Grasses for Dune Stabilization*, by Johanna Seltz, 1976. This brochure describes the importance of sand dunes and offers different means of stabilizing them through grass plantings. Available from UNC Sea Grant, North Carolina State University, Box 8065, Raleigh, NC 27695.

Flood insurance

50. *Questions and Answers/National Flood Insurance Program*, by the Federal Emergency Management Agency, 1983. This pamphlet explains the basics of flood insurance and provides addresses of FEMA offices. Available free of charge from the Federal Emergency Management Agency, Washington, DC 20472.

51. *Coastal Flood Hazards and the National Flood Insurance Program*, by H. Crane Miller, 1977. This 50-page publication describes in detail the nature of flood hazards in the coastal zone and the basic features of the flood insurance program. The most interesting portion of the guide is an attempt to assess the impact of the program on those interested in building in the coastal zone. Available from the National Flood Insurance Program, Federal Emergency Management Agency, Washington, DC 20472.

52. *Entering the Regular Program, No. 3*. This guide is intended for use by community officials during the period when a community enters the regular flood insurance assistance program. It explains the responsibilities of FEMA and the local community that must be met; also included is a timetable that should be followed. Available from the Federal Emergency Management Agency, Washington, DC 20472.

53. *Guide for Ordinance Development, No. 1e*. A guide designed for use by community officials in preparing floodplain management measures that satisfy the minimum standards of the national flood insurance program. It organizes the program's standards into a simple ordinance and provides an explanatory narrative. Available from the Federal Emergency Management Agency, Washington, DC 20472.

The next 6 items listed below (54–59) are available from the Federal Emergency Management Agency, Washington, DC 20472.

54. *Flood Insurance Study, Sarasota County, Florida*, by the National Oceanic and Atmospheric Administration, 1971.

55. *Individual County Flood Insurance Studies*, by the Federal Insurance Administration (FIA).

56. *Flood Insurance Study, Bay County, Florida*, by the FIA, 1981.

57. *Flood Insurance Study for Walton County, Florida*, by the FIA, 1977 (29 pp.).

58. *Flood Insurance, Escambia County, Florida*, by the FIA, 1977 (22 pp.).

59. *Flood Insurance Study, Santa Rosa County, Florida*, by the FIA, 1977 (25 pp.).

Construction, home improvement, and repair

60. *Elevated Residential Structures, Reducing Flood Damage Through Building Design: A Guide Manual*, by the Federal Insurance Administration, 1976. An excellent outline of the necessity for proper planning and construction in the face of the shoreline flood threat. Construction techniques are illustrated, and a glossary, references, and work sheets for estimating building costs are included. Order publication 0-222-193 from the Superintendent of Documents, U.S. Government Printing Office, Washington, DC 20402. Also available at offices of the Federal Emergency Management Agency.

61. *Design and Construction Manual for Residential Buildings in Coastal High Hazard Areas*, prepared by the Department of Housing and Urban Development, 1981. A guide to the coastal environment with recommendations on site selection and structure design relative to the National Flood Insurance Program. The report includes design considerations, examples, and construction costs, as well as appendixes on design tables, bracing, design work sheets, wood preservatives, and a listing of useful references. The manual is available from the Superintendent of Documents, U.S. Government Printing Office, Washington, DC 20402 (publication no. 722-978/545). Or you can find it at offices of the Federal Emergency Management Agency.

62. *Structural Failures: Modes, Causes, Responsibilities*, 1973. See especially the chapter entitled "Failure of Structures Due to Extreme Winds" (pp. 49-77). Available from the Research Council on Performance of Structures, American Society of Civil Engineers, 345 East 47th Street, New York, NY 10017.

63. *Flood Emergency and Residential Repair Handbook*, prepared by the National Association of Homebuilders Research Advisory Board of the National Academy of Science, 1980. A general guide to floodproofing, this handbook offers step-by-step cleanup procedures and repairs for household goods and appliances, among other items. Available from the Superintendent of Documents, U.S. Government Printing Office, Washington, DC 20402. Order stock no. 023-000-00552-2 for $3.50.

64. *Wind Resistant Design Concepts for Residences*, by D. B. Ward. Using vivid sketches and illustrations, this book describes construction problems and methods of tying structures to the ground. A considerable portion of the text and excellent illustrations are devoted to methods of strengthening residences. Recommendations are offered for relatively inexpensive modifications that will increase the safety of residences subject to severe winds. Chapter 8, "How to Calculate Wind Forces and Design Wind-Resistant Structures," should be of

particular interest to designers. Available as publication TR–83 from the Civil Defense Preparedness Agency, Department of Defense, the Pentagon, Washington, DC 20301, or from the Civil Defense Preparedness Agency, 2800 Eastern Boulevard, Baltimore, MD 21220.

65. *Hurricane-Resistant Construction for Homes*, by T. L. Walton, Jr., 1976. An excellent publication produced for residents of Florida, this booklet provides a good summary of hurricanes, storm surge, damage assessment, and guidelines for hurricane-resistant construction. Technical concepts on probability and its implications for home design in hazard areas are included. There also is a brief summary of federal and local guidelines. Available from Florida Sea Grant Publications, Florida Cooperative Extension Service, Marine Advisory Program, Coastal Engineering Library, University of Florida, Gainesville, FL 32611.

66. *Homes Can Resist Hurricanes*, by the U.S. Forest Service, 1965. An excellent guide with numerous details on general construction techniques. Pole-house construction is treated in particular detail. Available as research paper FPL 33 from the Forest Products Laboratory, U.S. Forest Service, P.O. Box 5130, Madison, WI 53705.

67. *Pole House Construction* and *Pole Building Design*. Available from the American Wood Preservers Institute, 1651 Old Meadows Road, McLean, VA 22101.

68. *Standard Details for One-Story Concrete Block Residences*, by the Masonry Institute of America. Written for both laymen and designers, this booklet contains 9 foldout drawings that illustrate the details of constructing concrete block homes. Principles of reinforcement and good connections aimed at design for seismic zones are discussed; these apply to design in hurricane zones as well. Available as publication 701 for $3.00 from the Masonry Institute of America, 2550 Beverly Boulevard, Los Angeles, CA 90057.

69. *Masonry Design Manual*, by the Masonry Institute of America. This 384-page manual covers all types of masonry construction, including brick, concrete block, glazed structural units, stone, and veneer. Comprehensive and well presented, it is probably of more interest to the designer than the layman. Available as publication 601 from the Masonry Institute of America, 2550 Beverly Boulevard, Los Angeles, CA 90057.

70. *Protecting Mobile Homes from High Winds*, prepared by the Civil Defense Preparedness Agency, 1974. An excellent booklet that outlines methods of tying down mobile homes and offers means of protection, such as positioning and wind breaks. Available free of charge as publication 1974-0-537-785 from the Superintendent of Documents, U.S. Government Printing Office, Washington, DC 20402, or from the U.S. Army, AG Publications Center, Civil Defense Branch, 2800 Eastern Boulevard (Middle River), Baltimore, MD 21220.

71. *Coastal Design: A Guide for Builders, Planners, and Home-owners*, 1983. See reference 3 of this appendix.

72. *Coastal Construction Practices*, by Christopher P. Jones and Leigh T. Johnson. This title includes an address list of state and local agencies involved in regulation of coastal construction. Available from the Marine Advisory Program, University of Florida, Gainesville, FL 32611.

73. *Guidelines for Beachfront Construction with Special Reference to the Coastal Construction Setback Line*, by Courtland Collier and others, 1977. Much to the distress of the authors of the present volume, the Sea Grant catalog says this publication illustrates criteria for evaluating variances that will uphold the purpose and philosophy of the coastal construction setback line wherever building seaward of the line is justified. Our distress stems from the fact that a setback line cannot have *any* exceptions if it is to succeed. At any rate, this publication can be obtained from the Marine Advisory Program, University of Florida, Gainesville, FL 32611.

74. *Building Construction on Shoreline Property*. A fact sheet on shoreline construction that is available from the Marine Advisory Program, University of Florida, Gainesville, FL 32611.

Building codes

75. *Coastal Construction Building Code Guidelines*, R. R. Clark, ed., 1980. This excellent volume, designed to supplement the 2 building codes listed below, is intended to strengthen these codes to meet structural design considerations required by section 161.053 of the Florida statutes. The *Standard Building Code*, 1979 edition, is available from the Southern Building Code Congress, 1116 Brown Marx Building, Birmingham, AL 35203, while the *South Florida Building Code* (*Broward County Edition*), 1979, is available from Broward County. A must for the conscientious builder, the coastal guidelines are available as technical report no. 80–1 from the Division of Beaches and Shores, Florida Department of Natural Resources, Tallahassee, FL 32304.

Index